"*On rainy days in the midst of haying season, work in the hayfield came to a halt. After we finished the morning chores and turned the cows out to pasture, we would crawl up into the haymow, where the freshly cut hay was stored. There we would rest on the hay that smelled of sweet clover and alfalfa and listen to the drumming of the raindrops on the barn roof. We'd listen to Pa's stories of rainy days he remembered. We enjoyed a day of rest and celebrated the rain, for our sandy farm never had enough.*"

Never Curse the Rain: A Farm Boy's Reflections on Water

*A companion to the*
*Wisconsin Public Television documentary*
JERRY APPS: NEVER CURSE THE RAIN

# NEVER CURSE THE RAIN

# NEVER CURSE THE RAIN

## A Farm Boy's Reflections on Water

Jerry Apps

WISCONSIN HISTORICAL SOCIETY PRESS

Published by the Wisconsin Historical Society Press
*Publishers since 1855*

The Wisconsin Historical Society helps people connect to the past by collecting,
preserving, and sharing stories. Founded in 1846, the Society is one of the nation's
finest historical institutions.
*Order books by phone toll free:* (888) 999-1669
*Order books online:* shop.wisconsinhistory.org
*Join the Wisconsin Historical Society:* wisconsinhistory.org/membership

Printed in Wisconsin, USA

Cover linoleum print by John Zimm © 2017
Design and cover illustration color by Nancy Warnecke, Moonlit Ink
21 20 19 18 17      1 2 3 4 5

Library of Congress Cataloging-in-Publication Data
Names: Apps, Jerold W., 1934–
Title: Never curse the rain : a farm boy's reflections on water / Jerry Apps.
Description: Madison : Wisconsin Historical Society Press, 2016.
Identifiers: LCCN 2016022476 (print) | LCCN 2016024268 (e-book) | ISBN
   9780870207945 (hardcover : alk. paper) | ISBN 9780870207952 (e-book) |
   ISBN 9780870207952 (E-book)
Subjects: LCSH: Water-supply, Rural—Wisconsin. | Rain and rainfall—
   Wisconsin. | Farm life—Wisconsin. | Country life—Wisconsin. | Water-supply,
   Agricultural—Wisconsin.
Classification: LCC TD927 .A67 2016 (print) | LCC TD927 (e-book) | DDC
   333.91009775—dc23
LC record available at https://lccn.loc.gov/2016022476

*If there is magic on this planet, it is contained in water.*

LOREN EISELEY

# Contents

*I dedicate this book to my late father, Herman Apps,
who helped me understand why we should never curse the rain.*

# Introduction

As a farm boy growing up on a hilly, sandy, stony, and droughty central Wisconsin farm, I learned early in my life to cherish water. The lives of our cattle and horses and of the crops that we grew depended on it.

We had three sources of water at our farm: the water we pumped from our deep well with a windmill, water that fell from the skies as rain, and water that fell as snow. None of these sources was dependable—the windmill sometimes didn't turn for days; the rains might come regularly one growing season and be spotty for another. Some winters the fields were covered with three feet or more of snow, and others the fields were as bare as a southwestern desert.

My dad instilled in me a great appreciation for the water we did have, and a bit of dread, too, for times when the rains didn't come, the weather was hot, and the windmill was still. When it rained, it was time for celebration. And when my brothers and I complained about a rainy day spoiling our plans, my father admonished, "Never curse the rain." Pa passed on to me an understanding of the great importance of water and how the farm's need for water must come before the family's hopes and wishes. Oftentimes this latter point was difficult for my brothers and me to understand, and I suspect it was even hard for my mother to accept at times. She never complained, and she cooked, cleaned, and washed clothes with as little water as possible.

Not all of our interest in water related to the farm's operation. My brothers and I enjoyed swimming in a nearby lake; we went fishing, more often in winter than in summer; we ice-skated; and everyone in our family enjoyed just sitting on the shore of a lake or a river and looking, listening, and thinking our own thoughts.

In this book I share some of my stories related to water. Some of the stories go back seventy-five years to when I was a youngster; others are more contemporary, as water has continued to be an important part of my life.

In a chapter I call "Final Thoughts," I write about my concerns for the future of water and the challenges we must overcome if we will continue to have sufficient and safe water for drinking, growing food, and enjoying our natural surroundings. Water is one of the most precious things on this planet, necessary for all life, and we must do everything we can to protect it through careful use, legislation, and regulation. For many of us this will require a change in our thinking. We can no longer assume we will always have the water we need.

# First Memories

By July of 1938, when my brothers were six months old, Donald was very ill. In fact, the doctor had told my parents he probably wouldn't live. None of us had been baptized—my dad didn't think it was necessary. But now my mother, who had grown up in a German Lutheran family and had attended a German school for her early schooling, insisted that Donald be baptized. Pa agreed. Although we didn't belong to a church at the time, we occasionally attended a Norwegian Lutheran church a few miles from our farm, and Ma asked the pastor, Reverend Carl Vevle, if he could come out to the farm and baptize the twins.

On a Wednesday evening, Reverend Vevle arrived at our farm. He asked for some warm water. Pa poured water from the pail next to the sink into a little pan and added warm water from the teakettle on the stove. My mother's brother, Uncle Wilbur Witt, and his wife, Katherine, had agreed to be witnesses for the event. I stood nearby watching, a curious little four-year-old (my birthday had been three days earlier) who had never seen anyone baptized. It seemed strange to me—and maybe even a waste of good water—as Reverend Vevle sprinkled a little water on Donald's head and intoned the words, "In the name of the Father, and of the Son, and of the Holy Ghost."

I immediately wondered which father he was talking about—my dad, maybe? And which son? And, wow, what was this holy ghost business? I knew enough about Halloween to have heard about ghosts, and now my sick baby brother was having water sprinkled on him in the name of a ghost, a holy one at that.

Darrel came next, the same sprinkling, same words. Ma watched carefully. Even at age four I could see all of this was very important to her. Darrel squirmed a little, but he didn't cry.

"Well, I guess that takes care of it," said Reverend Vevle as he began folding up the special little towels he had brought along.

"As long as you're here, Pastor, you might as well baptize Jerry as well."

"Sure," said Reverend Vevle. "I didn't know he wasn't baptized."

At my mother's direction, I bent my head over the pan. I got the full treatment, not just a sprinkling of water for me, but a goodly amount, as I heard the words, by that time committed to memory: "In the name of the Father, and the Son, and the Holy Ghost."

So now we were all three baptized—children of God, as the pastor said, patting me on the head. After Reverend Vevle left, I asked Ma what all of this baptism business was about, and she explained it to me as best she could. When I asked about the water, she said we had been baptized with Holy Water. It still seemed like ordinary water to me, having come from the water pail that sat every day on the edge of the sink. But I questioned her no further.

Donald's health improved (at this writing he is seventy-seven years old and lives just south of my farm in central Wisconsin). I've never forgotten the details of that day or the indelible role that water played in this important family event. It is one of my first memories of how important water was to our farm and our family.

# When the Windmill Quit Turning

On an unbearably hot August day in 1939, our windmill quit turning. The cattle tank that supplied water for our small herd of dairy cattle and our team of horses slowly emptied. I was only five, and until this day I hadn't realized how much our farm depended on a regular supply of water.

We were in the midst of a severe drought, as was much of the country at that time. Day after day the southwestern wind blew, bringing with it clouds of dust—dust that seeped into our house around the windows, dust that gathered on the oilcloth covering our kitchen table, dust that settled on newly washed clothes hanging on the clothesline so that they had to be washed again. But the wind also turned the windmill, pulling the pump rods up and down and bringing to the surface from some 180 feet deep cool, life-giving water.

Then the wind stopped and the temperature climbed even higher. The windmill didn't turn, the pump didn't move, and there was no water.

Pa watched the western sky, but there was not a hint

of rain on the horizon. The pasture grasses dried up and crunched underfoot. The little hay we had made in late June we now fed to the very hungry cows and horses. The situation became dire. Cattle will let you know when they are thirsty, and ours began bellowing, a loud, mournful sound that filled the dry, quiet air with a tale of misery. The cows pushed against each other at the water tank, now as dry as the barnyard. Dust hung above the herd in the still air.

Our farm community did not yet have electricity, and most of our neighbors pumped their water using windmills. The one exception was Alan Davis, who used a gasoline engine to power his water pump. On one of those hot, still days, I heard Pa on the party line telephone, ringing up Alan. He explained that our windmill had quit turning and we had no water for the cattle and the horses. He asked if we could come down to his farm with the team and fill a half-dozen ten-gallon milk cans with water.

Soon I was helping Pa harness the horses and hitching them to the wagon. We gathered up the milk cans and headed to the Davis farm, about a mile to the north on the same dusty country road that trailed by our farm. We could hear the bellowing of our cattle almost all the way there.

It took about an hour to fill the six cans with water from Alan's pump. As we waited for them to fill, we watered the horses. I listened as Pa and Alan talked about the drought and wondered when it would end. They talked about the Depression and how it had made life miserable for everyone, and how fortunate we were to live on farms and not in a city like Milwaukee or Chicago, where people might have water but had little else.

Back home, I helped Pa empty the cans of precious water into the stock tank as the cattle pushed and shoved and drank it almost as fast as we poured it. We saved enough to fill the water pail for our own use, about twelve quarts that would have to last us until the windmill began turning again or until we could make another trip to the Davis farm with our empty milk cans.

Then that night, as I lay in my bed upstairs, I heard the wind come up. I heard the squeaking and squawking of the old windmill as it began turning, once more pumping water and filling the stock tank. But the wind, at that moment our savior, was also the devil. The next day, clouds of dust filled the western sky, nearly blocking out the sun, leaving behind mounds of soil ripped from cornfields and grain fields, leaving behind dust like talcum that filtered into the house and covered everything—furniture, cookstove, icebox, and bedspreads.

Pa began scanning the for-sale ads in the *Waushara Argus,* looking for a used gasoline pump engine so that we wouldn't be without water when the windmill quit turning again. Soon he found one—a Monitor pump jack, a gasoline engine designed for pumping water. A farmer on the other side of Wild Rose had one for sale.

I remember the day the gasoline engine arrived on our farm. The windmill had stopped turning, and the water tank was once again empty. A truck backed up to the pump house, and Pa and the man selling the engine unloaded what looked like a huge hunk of cast iron. The pump jack consisted of gears, pulleys, and two arms that lifted and lowered a pump rod, much like our windmill did.

After a couple hours of shifting and pushing, bolting and pounding, the two men had the engine and pump jack in place. The engine stood about three feet tall, with a huge flywheel on one side and an open water-cooling tank on its top. A small tank for gasoline was fastened to one side of the engine—the first machine on our farm besides our car that required gasoline. (It would be several years before we had our first tractor.) Pa unhooked the pump rod from the windmill and fastened the rod to the new pump jack. He left the windmill standing.

The engine's former owner said we should dump a cup of oil into the machine every week or so to make sure the piston—which was about the size of two large fists put together—was lubricated. "Squirt a little oil on the moving parts every day, and it will serve you well."

He went on to demonstrate how to start the machine. "Here's what you do," he said, pointing to a wooden handle tucked into the huge cast iron wheel. "Grab hold of this handle. Before you start turning, push on this little spring. It releases the pressure from the cylinder so it will turn easier. When the flywheel starts spinning, let go of the spring."

I watched as Pa followed these directions and, with a bang and a shudder and a shake, brought the one-cylinder engine to life. It fired, and then it made all kinds of wheezing and puffing sounds before firing again. Meanwhile, the flywheel began turning. The three of us stood there marveling at the wonderful machine.

"Let it warm up for a couple minutes, then push on this lever and the pump jack will start," the man said.

Pa pushed on the lever, and the engine slowed some. The pops seemed louder and the vibrating a little more intense, but the pump jack began moving up and down. Soon water poured from the pump on its way to the water tank and the thirsty cows and horses.

Pa was smiling, and I was grinning, too. Now we could pump water when we wanted to, not when the wind allowed us to. When Ma heard the engine start, she came with the teakettle and water pail from the kitchen and returned with both filled. She, too, was smiling—the first smile I'd seen on her face in several days.

After a couple of hours the stock tank was filled with cool, fresh water. The animals drank, and the pushing and shoving ceased as the herd got its fill of water. They drifted off to where we had dumped some of our meager crop of hay, which was all they had to eat. Still, we knew they would survive the drought. And with the Monitor pump jack installed, one of Pa and Ma's major worries disappeared. As long as the well didn't go dry and the pump didn't break down, we would have water on our farm.

Some events one never forgets. This was one of them for me: the sound of thirsty livestock bellowing hour upon hour, the look of defeat on my father's face, and then the sound of that gasoline engine, with its wheezing and kabooming that resulted in water pouring into the stock tank in the barnyard.

# The Barn Came First

Pa had our water system set up so the water from the pump flowed first into a concrete cooling tank, which was large enough to hold six ten-gallon cans of milk. After every milking—sometimes two cans of milk resulted, sometimes three, depending on the time of the year—we immediately placed the cans in the cooling tank. We did this in all seasons of the year, as during the warmer seasons the tank cooled the milk, and during the winter it kept the milk from freezing. The milkman came by each morning to pick up the filled cans, leaving behind our empty ones.

An overflow pipe from the cooling tank thrust through a hole in the side of the pump house and led to the stock tank, located a few yards away just inside the barnyard fence. The cattle and horses drank from the stock tank year-round. In cold weather, Pa kept the stock tank from freezing by installing a wood-burning tank heater that was mostly submerged in the water. It was a small metal stove that had a little door a few inches above the water level where wood could be fed into it, and a stovepipe that extended several feet above the tank's water level. The stove created just enough heat to melt any ice that might accumulate on a cold winter night. Pa started the tank heater about the same time he went to the barn to do the morning milking. By the time he had finished milking and turned the cows out to drink at the tank, the ice had melted.

But on below-zero mornings, the cattle and horses were reluctant to leave the barn and trudge through knee-deep snow for their morning drink from the tank. Pa had been thinking hard about how to solve this problem. He was one to study the farm magazines, always looking for ways to improve his farming operation, and he knew that if the cows were more comfortable he could expect a bit more milk in the milk pail, and a little more money in his pocket. Then he read about a specially designed pump called a force pump that would not only bring water up from the well but would force it up a pipe as high as fifteen feet in the air.

By this time the start of World War II had brought higher milk prices, and Pa had a little money to invest in the farm. He bought some discarded telephone poles and enough two-inch iron pipe to cover the distance from the pump house to the barn. With the help of a local plumber, he installed a stock tank in the barn haymow, with water pipes leading from the tank to individual drinking cups in each cow stall. No longer would the cattle and horses have to walk outside in frigid weather to drink the cold water in the outdoor tank.

Now a cow could drink water whenever she wanted to. All she had to do was to push her nose against a metal flap, which opened a valve and allowed water to flow into the cup. Within a day or two, almost all of the cows had caught on. But one—I'll call her Florence, as all cows had names in those days—refused to use the drinking cup. In order for her to drink, Pa or I had to hold down the flap on her cup. After a couple of days of this, Pa and I went out to the barn and discovered a flood. The manure gutter behind the cows and the manger in front of them were full of water, and water was running out the back door. At first Pa thought a valve had stuck somewhere, or the water tank in the hay-mow had sprung a leak, but neither of those was the source of the problem. The problem was Florence. I watched as she pushed down on the drinking cup flap and the cup filled—and then as she kept pushing and it overflowed. She knew how to use the drinking cup, but she also seemed to enjoy pulling a watery trick on us. Pa put a stone under the flap of Florence's cup and removed it two or three times a day so she could drink. After a week of this, she caught on, shaped up, and began using the cup as it was designed.

Pa's setup for indoor plumbing in the barn was now running smoothly. One day I asked, "Pa, when do you think we'll have running water in the house?"

"Don't know," he said. "The money for this farm comes from the fields and the barn, not from the house."

It would be twenty years before Pa had indoor plumbing installed in our farmhouse.

# The Importance of a Well

The first settlers in the area generally sought farmland that either had access to a pond or a lake or had a stream running through it. Water was essential for farmers and their livestock, their crops, and their families—though generally the livestock and crops came first in the order of priority.

When a farmer had no access to a pond, lake, or stream, he had to turn to digging a well—not an easy task, but one people have taken on for thousands of years. The oldest known well was found in an archaeological excavation in Israel that was dated 8100–7500 BCE. Early wells followed the development of farming and the keeping of domestic animals. They were generally dug to water and lined with wood, not too different from how the early wells in this country were dug thousands of years later.

When someone asked Pa what was most important on the home farm, he never hesitated to answer: "A healthy family, fertile soil, a good herd of cattle, a trusting farm dog, a pair of willing horses, and a dependable well." A dependable well always appeared on the list, for Pa, along with every other farmer we knew, understood the importance of a ready supply of water.

During the early settlement days, which for our neck of the woods were the mid- to late 1800s, the wells were hand dug. On our home farm, where the well was 180 feet to water, stories passed on from owner to owner told of how the first settler here climbed into the hole he was digging, filled a pail with soil, used a pulley system and oxen to pull the pail to the surface, climbed back out to dump the pail in what later became our front yard, and then returned to the ever-deeper hole. The stories related that the hole had to be absolutely straight up and down. Any stones had to be removed and the sides of the hole cribbed up with wooden planks so the soil would not cave in on the digger.

For the well on our home farm, one of the early owners placed a two-inch pipe from top to bottom in the open hole. He shoveled some of the soil back in the hole, filling it to within eight feet or so of the surface, thus preventing vermin and trash from falling into the open well and contaminating the water. The pump rod from the windmill-powered pump moved up and down within the two-inch pipe, bringing the water to the surface.

For a well to be dependable, it had to be deep enough in the aquifer to assure a steady supply of water even during drought cycles. This meant that the well also must be reasonably easy to dig. Our neighbor Alan Davis moved his entire farmstead a half-mile from its original location because stony soil prevented him from digging deep enough to reach the aquifer.

John Coombes, who once owned the land I own now, did something similar. In 1912 Coombes moved his entire farmstead a few hundred yards to the north, to the other side of the town road that divided his property. He constructed a new well and installed a windmill-powered pump. And this time he had insurance for when the wind didn't blow and his pump didn't work. The part of his farm that lay north of the road, some twenty acres, included a small pond. When he built his barnyard fence, he also fenced a quarter-mile lane from his buildings to the pond so his cattle and hogs and horses could walk to a water source when the windmill stopped turning and the stock tank was empty.

For thousands of years, most farmers depended on wells for their water. This dependence continues to this day, although technology has made well drilling much easier than in the days when a man with a shovel dug the well hole.

# Saturday Baths, Monday Washing

When asked if there was running water in our farmhouse, Pa's ready reply was, "Sure—grab a pail, take it to the pump, fill it up, and run back to the house." Then he would throw back his head and laugh as if he had made the biggest joke imaginable. But it was no joke for my mother, brothers, and me. As long as I lived at home, we had no indoor plumbing. We did have a party-line telephone, and electricity came our way in 1947. But we would not enjoy the benefits of a flush toilet, a shower and bathtub, and a kitchen sink with hot and cold running water until 1955.

How did we manage? Or as one youngster once asked me, "How did you survive?" (I reminded him that for thousands of years, people had no electricity or indoor plumbing—I guessed he must have snoozed through those history lessons in school.)

With no indoor plumbing, my brothers and I learned the value of water without anyone having to tell us. We learned it by having to carry pail after pail from the pump to the house, where we filled the teakettle and the cookstove reservoir and then set the once-more filled pail on the counter near the kitchen sink. That would be our mother's water for cooking and making coffee, our drinking water, and our water for brushing teeth and washing our dirty hands and faces after a day's work in a dusty field.

Two times during the week, we carried more pails of water than usual—for Saturday night baths and for Monday morning clothes washing. Bath night began after the cows were milked and turned out to pasture and the cans of fresh milk were dropped into the cooling tank in the pump house. Pa retrieved the large galvanized metal washtub, which hung on a nail in the woodshed the rest of the week. (When one of my mother's city relatives would ask, "You mean you bathe only once a week?" Pa would reply very seriously, as if imparting a vital piece of rural wisdom, "Don't you know that too much bathing will weaken you?") My brothers and I carried several pails of water to the house, enough to fill the teakettle and the cookstove reservoir and enough to fill the washtub about half full, which Pa declared was enough for a decent bath.

After the evening milking, we filed back into the house, where Ma had a teakettle steaming on the woodstove. The water in the cookstove reservoir was also warm and ready, as Ma had kept the stove going after cooking the evening meal.

In those days the soap of choice was Lifebuoy, a carbolic soap that smelled of disinfectant. We had a big bar of it, and we all used it. We took turns bathing, my brothers first, then me, then the folks, all using the same water, with a little more warm water added as it cooled. When the bathing was completed, and we all smelled fresh and clean and "disinfected," we pulled on clean underwear, clean bib overalls, and clean blue work shirts. Pa dumped the washtub water out the back door and hung the washtub on its nail in the woodshed, and we piled into our 1936 Plymouth for our trip to town. Saturday night bath night was also town night.

The second time during the week when we toted several additional pails of water from pump house to kitchen was Monday morning, which was washday, no matter if it was midsummer and the temperature hung in the high eighties, or it was winter and it was thirty below zero.

For six days of the week, a copper boiler with cover sat unused behind the kitchen cookstove. On washday, it found its place over the hottest part of the stove, to be filled with several pails of three-sons-toted water. Not only did we carry water to fill the copper boiler, we also carried enough to fill the two rinse tubs. Each rinse tub held ten gallons of water; the washing machine needed fourteen gallons (which we put in the copper boiler to heat). So altogether we carried into the house thirty-four gallons of water for washday. With a water pail's capacity of about three gallons, we carried about twelve pails of water from the pump house to the house—two pails at a time, so it took six trips.

One of our modern conveniences included a Briggs and Stratton–powered Speed Queen washing machine, which Pa bought in the early 1940s. No more scrubbing dirty clothes by hand with a washboard—although Ma still used a washboard on occasion when our bib overalls had come in contact with some dirty mess that she knew the Speed Queen did not have the gumption to remove. On washday Pa pushed the kitchen table out of the way and set up the washing machine in front of the kitchen stove. The two washtubs were pushed up tight to the washing machine's wringer, to keep water from spilling on the floor when clothes were run through the wringer.

The Speed Queen had one problem, a problem that—and Pa didn't realize it at the time—provided an opportunity for his sons to broaden their vocabularies, especially in the category of cuss words. The theory for starting the Speed Queen engine was simple: pull up on the choke wire and, with one foot, push down on the kick-starter, and it should take off. But nine out of nine Monday mornings, the Speed Queen didn't start on the first kick, or on the tenth kick. With every kick after the first five or six, the air became filled with Pa's cuss words, until finally, with a pop and a belch of black smoke, the machine would start to run and washday would begin.

With the Speed Queen purring—I'm sure it sounded like purring to Ma and Pa, but it was really quite an obnoxious noise—Ma added hot water to the water in the machine, which by this time had already cooled. With the machine going, Ma shaved homemade laundry soap into the water, creating a little mountain of foamy suds.

She began with the underwear, towels, bedsheets, dresses, and aprons—the white things. After washing those items, she ran them through the ringer into the first washtub to remove the soap. Then she ran them through the wringer into the second washtub, and once more through the wringer into the clothes basket.

After adding more warm water from the woodstove's reservoir, Ma tossed the second load, which included shirts and socks, into the washing machine. They went through the same cycles as the white things. She added more warm water and then tossed the third load into the washing machine: the bib overalls, usually the dirtiest of all the clothes to be washed.

After all the clothes had been washed, rinsed, and run through the wringer, Ma carried the wet items out to the wire clotheslines strung between two posts a short hike from the back door. There she hung the clothes to dry, no matter what the temperature. It was not a pleasant job. By midafternoon she would bring the now dry clothes back into the house in preparation for ironing, which she did on Tuesday of each week.

After washday, we went back to the typical routine of carrying in the two or three pails necessary for the daily operation of the kitchen—until Saturday bath night came around again.

Occasionally we were reminded that those who had indoor plumbing used water far more freely than we did. For two weeks every July, George and Mable, friends of my parents, visited from Chicago. George was content to sit under a shade tree and read the day-old *Milwaukee Journal* that was delivered to our mailbox. But Mable was intent on helping Ma in the kitchen. And when Mable was in the kitchen, she could never seem to have enough water. "Jerry [or Donald, or Darrel], dear, would you fetch me another pail of water?" was her never-ending call. My brothers and I invented elaborate strategies to be missing at these times. When we weren't working in the hayfield, we hid in the woods, the barn, or another of our assorted hiding places. Even Pa, who generally avoided criticizing people, noted, "That Mable sure uses a lot of water."

# Wild Rose Gristmill

Just as water was important to individual farmers and their operations, it was vital to the life of our entire community. During the Midwest's early settlement days, many villages and cities located along a lake, river, or stream. Easily accessible water provided for many of the needs of the local citizens, including power for gristmills and sawmills. In 1882 a *Waushara Argus* reporter wrote about my hometown, "The Village of Wild Rose is very pleasantly situated on a branch of the same stream that passes through Saxeville, Pine River and Poy Sippi. It takes its rise in the northern part of Rose Township and empties into Lake Poygan. In this place there is quite a fine waterpower that runs a gristmill, and a sawmill has been put in and considerable small timber is sawed up."

From the 1930s through the 1950s, the community surrounding the village of Wild Rose consisted of small dairy farms. Before electricity came to the area (not until after World War II for most), farmers fed and milked their cows by hand by the light of kerosene lanterns. They grew almost all of their own cattle feed, which consisted of hay, stored loose in the haymow of the cattle barn; silage, made from corn that was cut into small pieces and fed into an upright silo; and grist, generally a combination of ear corn and homegrown oats that were ground together at the gristmill in Wild Rose.

When I was three or four years old, I remember riding with Pa to the old water-powered gristmill. Pa filled several sacks full of ear corn from the corncrib and filled several more with oats, then piled all of the sacks into the back of our 1936 Plymouth with the back cushions removed. On the four-mile ride from our farm to Wild Rose, the heavily laden Plymouth moved at a top speed of thirty-five miles an hour, squatting the entire way, its front end pointing up a few degrees from normal.

Once we arrived at the mill, we often had to wait an hour or more while other farmers unloaded their corn and oats on the mill porch and waited their turn for grinding. Then we'd wait some more while our grain was ground. In the summer, Pa and I sometimes sat on the banks of the millpond in the shade of a weeping willow tree. We'd talk about whether the trout that lived in the pond might be biting and how some rainy morning when we couldn't do farmwork we ought to go there with our fishing poles and try our luck.

More often Pa struck up a conversation with one of the other farmers. Weather was the main topic; it was either too wet or too dry in the minds of the farmers, most often too dry, for our sandy soils required considerably more rain than the richer soils in southern Wisconsin.

We both made sure to get a drink from the always-running water pipe near the entrance to the mill. The pipe led to a spring in the side of the hill near the millpond, and from it poured the clearest, coolest, best-tasting water, ranking right up there with the well water from our own well. In winter we'd crowd into the miller's office, a tiny space warmed by a potbellied wood-burning stove. The main feature of the dusty but warm little room was a massive rolltop desk, the kind with many little drawers. Stacks of papers were piled everywhere, all covered with a fine dust from the mill. Inside the office the discussion was again about the weather. Either it was too warm, and the snow cover was melting, exposing dormant hay, or it was too cold and simply unbearable to be outside. Farmers had more time in winter than in summer, so these conversations were longer and more relaxed. In addition to the weather, they talked about their cows, crops, fishing, and politics. These were the Depression years, so a lot of the talk, summer and winter, was about the state of the economy. Farmers speculated whether farm prices would ever rise to a level that would allow them to have something left after paying the bills and the interest on the farm mortgage.

Being a curious kid, I often wandered around the mill watching the belts and pulleys, listening to the hum of the mill machinery grinding the oats and corn, and feeling the huge old timbered building shake and shudder as the machinery worked. The mill was powered by a turbine housed in a steel encasement. I could see where the water from the millpond flowed through the penstock to the turbine. And I could see where the water left the turbine on its way back to the river. But the turbine itself was hidden from view.

On the mill floor, where the grain was received, I could see none of the machinery. The miller absolutely forbid me or anyone else to follow the stairs to the lower reaches of the mill where the grinding equipment was located; only the miller got to go there.

When it was our turn, Pa untied the binding strings from the corn and oat sacks and dumped the contents into little square holes in the floor, where it rattled its way to the basement of the building and to the grinding machinery. The cob corn and the oats were mixed together during the grinding process, which produced a warm, fluffy, sweet-smelling grist that our cattle relished. After the grain was ground, it was elevated to the floor above and then fed by gravity into the same bags in which we'd delivered the corn and oats. The sacks of grist were dragged onto the huge scale near the sacker and weighed, and we paid the miller according to the pounds of grist ground. Then we loaded it into the back of the old Plymouth and hauled it home. We repeated this about once a week throughout the year.

Like many little mills all over Wisconsin, the Wild Rose gristmill relied on waterpower to do more than grind grain. It also generated electricity.

When I was growing up, living conditions in the country were considerably different from those in the villages and cities. Most people in town had central heating, with a furnace in their basement. Farm people heated their homes and did their cooking and baking with wood-burning stoves. Most town people had indoor plumbing, while most farm people did not. But what most profoundly separated town folks from farm folks was electricity. We lived with lamps and lanterns to light our way; we pumped water with windmills and gasoline engines. We cooled our food with an icebox, and Ma washed clothes with a gasoline-powered washing machine and ironed clothes with a sad iron warmed on a wood-burning cookstove.

As a farm kid without electricity, I was ashamed to invite my friends from town to visit our farm. If you had electricity and had become accustomed to its many advantages, I figured you likely didn't want to sit in a dim room lighted with a kerosene lantern or drink lukewarm milk from an icebox. And, horror of horrors, who would want to visit a frigid outhouse on a winter evening when he was accustomed to indoor plumbing? Waterpower helped create this gap between life in the country and life in the villages and cities.

In 1908 miller Ed Hoaglin brought electricity to Wild Rose. He built a little brick building just outside the gristmill, and in it he installed an electric generator. Hoaglin and his hired man strung electric wires around the village of Wild Rose so that the homes and businesses could have electricity. I'm sure folks paid Hoaglin and his men for the work, but I have no record of how much it was.

With the water-powered generator creating electricity, villagers replaced their kerosene lamps with lightbulbs, clothes washtubs with electric washing machines, and sad irons with electric irons. But one obstacle to convenience remained. Because the mill was busy grinding grain all day long, requiring a substantial amount of waterpower to operate, no electricity was available until after five p.m., when the grinding operation at the mill shut down. The electricity was turned off once more at eleven p.m. each night, so the millpond could build up a sufficient head of water to power the mill the following day. The mill also shut down each day at noon for an hour, while the milling crew ate lunch.

In exasperation, a group of village women that included the miller's wife marched to the mill one noon and confronted the miller. They explained that they wanted the electricity on so they could iron clothes in broad daylight. The miller relented and turned on the electricity in the village for an hour. While the women of the village quickly washed and ironed clothes, the men at the mill ate a leisurely lunch and rested until one p.m., when the miller once more shut off the electricity and resumed grinding grain.

Those of us living on farms had little sympathy for the women in Wild Rose and their demands for more time to use the electricity that was available in their homes. Electricity did not come to our farm until 1947, nearly forty years after the town folks had lightbulbs and electric irons. By this time, people in the village were getting electricity from a private electric company, one that was reluctant to string lines to farms. The Rural Electrification Administration (REA) of 1936, one of President Roosevelt's New Deal programs, was a major force in bringing electricity to the country. But World War II stopped the march of REA electric lines into the country, and we had to wait until war's end to have electricity on our farm.

I have not forgotten the years when that great divide existed between my friends living in Wild Rose and those of us who lived on farms. For all that water brought to our lives, in this case water caused a rift.

# Millpond Memories

The Wild Rose millpond provided waterpower for a grist-mill and later provided electricity for the village. But the millpond had much more to offer. We could catch a brook trout there with a cane pole, a hook and line, and a wiggly worm. The local kids skated on its frozen surface in the winter. When the ice was thickest, usually in February, the two icehouse owners and their crew of men cut the mill-pond ice into huge chunks and stored the ice in their ice-houses, covering it with sawdust to keep it from melting. During the hot days of summer, the iceman made the rounds of the neighborhood, filling customers' iceboxes with fifty-pound hunks of millpond ice.

Leading up from the millpond was a long, sloping hill-side where wooden planks had been fastened to posts driven in the ground—a natural amphitheater—and on Tuesday evenings in late May through early September, we watched free movies there sponsored by the village's merchants. The movie screen was a bedsheet fastened to a two-by-four piece of lumber, which in turn was nailed to an enormous willow tree on the edge of the millpond, its long bundles of leaves hanging over the water.

A summer Tuesday night was special, especially for my brothers and me. On "free show night" we hurried to finish the evening milking, turn the cows out to pasture, and then ride with our parents into Wild Rose in plenty of time to see the movie, which started after the sun had set.

Upon arriving in town, we found a place on one of the plank benches. They quickly filled with people from the village and with those living as far as ten miles away. We knew the movie was about to start when "Specks" Murty, the village marshal, arrived. With a rope he let down the streetlight on Main Street that was adjacent to the free show area, screwed loose the bulb, and pulled the now dark streetlight back in place.

As darkness engulfed us and steam began rising from the millpond water as the cool night air met the warm water, a whippoorwill began calling its name: *"Whippoorwill... whippoorwill... whippoorwill."* A bullfrog joined the chorus of the night sounds with its deep *"Harrumph, harrumph, harrumph."*

The projectionist turned on the movie machine (which is what we called it), and the movie began. The movie soundtrack competed with the more primitive sounds of the night—the calling of the whippoorwills and the bullfrogs and the occasional splash as a trout surfaced to gather up an insect. To me the night sounds were as interesting as the movie. Those sounds emanating from the millpond during free show night are forever imprinted in my memory.

29

# Chain O' Lake

Chain O' Lake was a small, shallow lake about a mile and a half from our farm. Our community, our one-room country school, and the 4-H club we organized in 1945 were all named for it.

People regularly tried to correct me when they asked what school I attended, or what 4-H Club I belonged to, and I answered "Chain O' Lake."

"You mean Chain O' *Lakes*," they said. "There can't be a 'Chain O' Lake' if there's but one lake. It doesn't make sense! *Chain* suggests several lakes tied together."

Several small lakes do exist in this mile-long valley where Chain O' Lake (about twenty acres) is the first in line and Wautoma Lake (about eleven acres) is last. But although they are relatively close together, they are not "chained" together like the Chain O' Lakes at Waupaca, where twenty-two lakes of various sizes, shapes, and depths are stitched together so a boat can move from one to another with little difficulty.

Someone who seemed to know how underground water systems work once told me that these lakes beginning with

Chain O' Lake are in fact connected, but they are linked underground. That seems to make sense, for the water level of these lakes (some little more than ponds) do rise and fall together. When the water level of Chain O' Lake is down, so is the water level on all the other lakes and ponds in the system.

None of these ponds has an inlet or an outlet like many lakes in the area do. They are water table lakes, which means that their water levels change depending on the water table. Evidence suggests that when the glacier receded from this area some ten thousand years ago, huge chunks of ice remained embedded in the soil. When the ice eventually melted, the lakes were formed.

In the early 1930s, Chain O' Lake, already a vital source of water for this farming community, became important to local farmers in another way. Like several other lakes in the area, Chain O' Lake had embedded in its bottom a rich deposit of marl. Marl, formed thousands of years ago by the decaying of an aquatic plant called chara, is rich in calcium carbonate. It proved to be an important new idea for the farmers trying to grow enough feed for their small dairy herds in western Waushara County. Researchers at the University of Wisconsin's College of Agriculture and its field-based county extension agents were promoting the use of a new high-yielding forage plant, alfalfa. My grandfather William Witt heard about alfalfa and quickly decided he would like to grow some on his sandy farm a mile west of our home farm. Grandpa Witt also learned that alfalfa required alkaline soil (or "sweet soil," as some described it).

The soils in our community were acidic, anything but sweet. The county agents suggested that farmers in our area could "sweeten" the soil by applying calcium carbonate–rich marl. The University of Wisconsin Agricultural Experiment Station in 1896 published a bulletin titled *The Marls of Wisconsin,* by F. W. Woll, extolling the virtues of marl as a soil sweetener.

Grandpa Witt enlisted the cooperation of his neighbors, and together they gathered the funds to hire a steam-powered dredge to dig the marl from the bottom of Chain O' Lake, some fifteen feet down, and dump it on shore in a huge pile. After allowing the marl to dry for several months, the farmers hauled horse-drawn wagon load after wagon load of it to their sandy fields. They spread it over the soil by hand and then plowed it down and planted alfalfa. They saw excellent results. I once overheard Grandpa Witt telling my father until he spread marl, all his attempts to grow alfalfa had failed. With marl he had gotten several loads of alfalfa hay from a ten-acre hay field.

The marl-digging project provided another benefit: about fifty yards from the shore of Chain O' Lake, the dredging created an extra-deep spot where on the hot summer days the big bluegills, sunfish, and perch gathered.

During the war years, Pa often hired a man for the summer months. Usually young and single, our hired hand would live with us as a part of our family from about April until September. One of these young men was Henry Lackelt, who was tall and thin and in his early twenties. Somehow he had avoided being drafted into the army. I never thought to ask him; perhaps he was 4F due to some medical condition. Henry enjoyed farmwork, never complained, and liked to fish. He also owned a Model T Ford touring car with a cloth top that could be removed; today we would call it a convertible.

Oftentimes in late June and July, on a warm summer evening when the chores were done and the cows were all out to pasture, Henry, my brothers, and I would dig a few earthworms in the rich soil in back of the chicken house, gather up our seventeen-foot cane poles stored under the corncrib eaves, and chug off to Chain O' Lake in Henry's Model T. Henry backed the Model T into the lake as far as he dared without getting stuck, and we'd all fish from the back of it.

The fish didn't always bite, but we expected that, and we brought along our bathing suits. With no fishing action, we'd wrap up our cane poles, change into our swimsuits, and leap off the back of the Model T into the cool waters of Chain O' Lake. As darkness slowly moved across the lake, Henry would say, "We'd probably better head for home." He would crank up the Model T (it had no fancy electric starter), and my brothers and I would jump into the lake and push the car toward shore. Soon we were once again driving along the dusty country road, enjoying the sweet smell of summer and feeling the cool breeze of early evening sweeping across our faces. I've never forgotten those fishing-swimming nights in Chain O' Lake. What fun they were, and how fortunate we were to have a lake so close to our farm.

# One-Room School

When I enrolled in Chain O' Lake School in 1939 at age five, the little building had no electricity, was heated with a woodstove, and depended on a hand pump for water. A boys' outhouse stood on the southwest corner of the school-yard and a girls' on the northwest corner.

The number of students attending Chain O' Lake from year to year ranged from fifteen to twenty-five in eight grades. We each had chores to do. Some of us carried in kindling from the woodshed a hundred feet or so from the school building (I had that job in first grade), others swept the outhouses each morning, and still others were re-sponsible for cleaning the blackboards that filled the wall on the west end of the building. Each morning students in fourth through eighth grades took turns carrying hunks of oak wood into the school and piling them in the entry-way for the ever-hungry woodstove that sat in the back of the room. And no matter if it was eighty degrees or twenty below zero, a student was assigned to pump and carry in enough water to fill the five-gallon Red Wing water cooler, which sat next to the sink in the back of the room. A pipe led from the sink and spilled the used water out on the ground alongside the school building. The water cooler was our sole source of water—for drinking, for hand washing, and for occasionally filling the pan on top of the wood-stove, which added a little humidity to the room and in winter heated our jars of homemade soup. It was fancier than the water pail in our kitchen at home, where we drank from a dipper. The Red Wing cooler had a shiny, curved metal tube that extended from the bottom and ended at a little metal bowl hanging over the sink. When you pushed a button, water burst up from the bowl a couple of inches or more, depending on how much water there was in the cooler. A wash pan stood under the little metal bowl and was used for hand washing before lunch and after visiting an outhouse.

One of the dirty tricks older students played on incoming first graders involved the water cooler. I remember an older student—probably Clair Jenks, who was in fourth grade—asking me, "Do you know how to get a drink from the water cooler?"

I answered that I did not; indeed, I had never seen a water cooler before and didn't have the first idea about how it worked. At home I simply dipped our dipper into the open pail to get a drink.

"Here, I'll show you," Clair said. "First you lean over this little bowl in front of the cooler and get your nose as close as you can to this hole." He pointed to the little hole at the end of the pipe.

When my nose was but an inch or so from the little hole, Clair pushed the button and a stream of water shot into my nose. I jumped back in surprise, water running down my face and dripping on the floor while Clair stood by laughing. After that I knew that to get a drink, I should carefully hold my mouth a few inches from the hole while I pushed the button.

Another duty for the student responsible for hand pumping water and filling the Red Wing cooler was to make sure that the little metal pipe was thawed out on cold winter mornings. During the night the woodstove that heated the building went out, and if the temperature outside skidded well below zero, as it often did on winter days, the temperature in the schoolroom fell to near zero, freezing what water may have remained in the pipe. Thawing it was easy, although it was not an especially pleasant task. The assigned student had to grab hold of the frozen metal tube and hold it tight until it thawed. It took sometimes ten or fifteen minutes, but the warm hand on the frozen tube worked every time.

On the last Friday in April we celebrated Arbor Day at our country school. Arbor Day is associated with planting trees, but our schoolyard had about as many trees as could be accommodated on the acre lot, so we spent the day raking the lawn, picking up dead branches, and making the schoolyard neat and tidy. With everyone pitching in, we usually had the raking and cleanup finished by noon. In the afternoon, all the students and our teacher walked the half-mile to Chain O' Lake. We walked all the way around the lake, looking for frogs and turtles and peering into the water for signs of fish, watching wild ducks fly up, and listening to the red-winged blackbirds that sat on last year's cattails, singing their hearts out in celebration of spring.

Once back at the school, we gathered around the huge pile of leaves we had raked, and our teacher touched a match to the pile. When it had burned down a bit, we enjoyed roasted wieners and marshmallows and talked about what we had seen, smelled, and heard as we walked around the lake. It was a wonderful day that we looked forward to every year—a celebration of spring after a long, snowy winter.

# Rainy Day in June

I remember one June morning many years ago. When I woke up at five-thirty, I heard the splash of raindrops on the window of my second-floor bedroom, and I saw the water trickle down to the windowsill. The morning was gray and dreary. The rain had begun with no announcement. No booming thunder. No flashes of lightning. No tree-shaking wind.

I dressed and made my way downstairs, where I pulled on an old tattered raincoat and my equally well-worn barn cap. Soon I was slogging my way along the half-mile lane to where the milk cows were waiting for me in the night pasture. Fanny, our farm dog, accompanied me, seeming to enjoy the gentle rain that was turning the once dusty cattle trail into mud. I plodded along, my head down, with rainwater dripping from my cap and running from the bottom of my raincoat.

When we arrived at the night pasture, Fanny rounded up the cows and young stock with a bark or two, and they began the half-mile walk home, in single file, the rain dripping from their black and white coats. The rain had become heavier, but the cows seemed not to be bothered.

Pa opened the barn door when we arrived, and the milk cows filed in, taking their places, always the same stalls, as they had been trained to do. The young stock continued on past the barn door. They would wait in the barnyard until the cows were milked, and then the entire herd would walk back to the pasture, where they would spend the day grazing, chewing their cuds, and resting.

The rain increased in intensity. Not a downpour, but a steady, soaking, earth-loving rain. It rained all day and all night and most of the following morning.

On rainy days like this in the midst of haying season, work in the hayfield came to a halt. After we finished the morning chores and turned the cows out to pasture, we would crawl up into the haymow, where the freshly cut hay was stored. There we would rest on the hay that smelled of sweet clover and alfalfa and listen to the drumming of the raindrops on the barn roof. We'd listen to Pa's stories of rainy days he remembered. We enjoyed a day of rest and celebrated the rain, for our sandy farm never had enough.

Sometimes when the rain stopped but the clouds remained and it was too wet to work in the fields (even too wet to hoe potatoes, which was the fallback job when all other work was finished), Pa suggested we go fishing. Farmwork always took precedence over fun activities, so fishing in the summer was a rare treat.

Our seventeen-foot-long cane fishing poles, with heavy green fish line wound around them, hung under the corncrib eaves. My brothers and I quickly gathered them up and then grabbed a six-tine barn fork and dug a can of earthworms from the rich soil in back of the chicken house. With binding twine, we tied the cane poles over the top of our old 1936 Plymouth, then grabbed up some spare hooks purchased at Hotz Hardware in Wild Rose and crawled into the Plymouth.

Our destination was Norwegian Lake, a great place to catch bluegills, the occasional black bass, and, if we were lucky, a northern pike. There was no public access to Norwegian Lake, but Pa knew the Andersons, whose farm abutted most of one shore and whose short wooden pier jutted out into the lake. Tied to the pier were three wooden boats that the Andersons rented out for a dollar whether you fished for an hour or you fished all day. Stopping in at their house, Pa chatted for a bit with Millie Anderson. Her sister, Maxine, was my teacher at Chain O' Lake School. Pa handed Millie a dollar, walked to their wagon shed, and gathered up a pair of wooden boat oars and stuffed them into the backseat of the car. He drove along the narrow, rutted wagon trail a half-mile or so to the pier, where we found the boats, each containing an ample amount of rainwater from the recent storm. The boats were badly worn after years of fishing and little care. Pa selected one of the better-looking ones.

My brothers and I began bailing water from the boat with an old empty coffee can left for that purpose. Pa untied the poles from the Plymouth, grabbed the can of worms, and put everything in the boat, whose bottom was now mostly dry. He put the oars into the oarlocks, and we pushed off from the pier. (We had no life jackets and didn't even know such things existed.) Pa began rowing toward the old marl hole, where marl had been dredged from the bottom of the lake. The hole was the deepest part of Norwegian Lake; Pa figured it to be thirty feet, maybe more, and the deeper water remained at a constant cool temperature. We knew fish gathered there.

After about fifteen minutes of rowing, I watched the bottom of the lake disappear from sight, and I knew we had arrived at the marl hole. Our anchor was an old iron plow point no longer serviceable for plowing but heavy enough to hold a boat in place. The rope attached to the anchor wasn't long enough to reach the bottom of the marl hole, so we knew we had to anchor the boat on the edge of the deep hole but close enough that we could toss our lines into the deep water. It took a bit of maneuvering to find just the right place. I held the anchor at the ready, my brothers stared at the bottom, and Pa slowly rowed. Finally, the twins shouted, "Now!" and I dropped the anchor with a splash. We watched it sink to the bottom, with several feet of rope to spare.

"The marl hole is right there," Pa said, pointing to a place to the right of the boat. A slight breeze blew from the west, moving away the last of the rain clouds. With the appearance of the sun, I immediately thought that our fishing trip would be a short one, as I was sure Pa was thinking of some task that needed doing at the farm now that the sun was out. But he said nothing as he unwound the thick green line from his long cane pole. My brothers and I did the same thing. We attached bobbers to each of the fish lines, and Pa suggested where we should place them so our worm-baited hooks would hang about six feet below the bobbers.

Soon four red and white bobbers danced on the lake surface, which was slightly choppy now as the breeze grew a bit stronger. We stared at our dancing bobbers, and at one another's. We didn't say much, but for my brothers and me, fishing, like so many things we did in those days, was competitive. Who would catch the first fish? Who would catch the biggest fish? Who would catch the most fish? Winning in one or more of these categories gave one a storytelling edge when neighbors saw each other and the question came up, "Been fishing lately?"

Donald's bobber went under first. It disappeared, appeared on the surface, then disappeared again. Pa said, "Set the hook and pull him in." Donald lifted the long cane pole, tightening the fish line. As he lifted the pole, it began bending, a sure sign that a fish was on, and a good-sized one at that. "You got a pole bender," Pa said.

Soon a bluegill emerged from the water, a fish about the size of Pa's open hand, larger than the average ones that we usually caught. And then Darrel's bobber went under, and then Pa's, and then mine. As we caught each fish, we unhooked it from the line and tossed it into a pail half-filled with water. Each time there was considerable splashing and thrashing as the caught fish attempted an escape.

I noticed it first. The bottom of the boat was covered with water. I wondered if we hadn't bailed out all the rainwater. But that wasn't the case. We had been so busy catching bluegills that we hadn't noticed that the old wooden boat was leaking badly. We had no life vests, no boat cushions, nothing to help us if the old boat sank. The water was beginning to rise up the soles of our shoes, and the boat's sinking seemed a clear possibility.

"Pa, the boat is leaking," I said.

"I know," Pa said, not sounding too worried. "Take turns bailing. Fish are biting, makes no sense to leave."

I wanted to mention that the boat might sink if we couldn't keep up with the bailing. But I didn't. If Pa thought the boat wouldn't sink, then it probably wouldn't.

My brothers and I bailed water. When we weren't tossing water out of the boat, we were pulling fish into the boat. It was one of the better fishing days that I remember. I noticed, too, that our shoes were getting wetter. Water was coming into the boat faster than we could toss it out. In between catching fish and rebaiting his hook, I saw Pa glance down to see the water now halfway up his shoes.

Finally, about the time I was thinking that the boat might indeed sink, and in water well over our heads, Pa said, "Boys, we might as well roll up our lines and head to shore. We've got enough fish for a couple of meals."

Without even mentioning the water now sloshing around in the bottom of the boat, Pa rowed to shore. We tied the sinking boat to the Anderson pier, fastened the fish poles to the top of the Plymouth, put the pail of fish in back, and headed for home. It had been a great day of fishing, one that left us full of stories to be told about big fish and little fish and especially lots of fish. None of us mentioned that our boat leaked like a sieve, especially to Ma, who would not have been happy. Instead we presented her with our catch, and she fried them up that very evening. Pa often said that nothing tasted better than fresh-caught skillet-fried bluegills.

# Haying Season

Pa had a litany of sayings about rain: "April showers bring May flowers." "Rain in May is a barn full of hay; rain in June is a silver spoon." "Rain in July is a welcome rain." April and May rains were critical to a decent hay crop. Some years the rains were ample during these months, but in some years they weren't, especially during the Depression of the 1930s when a widespread drought settled over much of the country.

During the war years of the early 1940s, the rains came again and the hay crop flourished. After many long days of hard work, our barn haymow was filled to the rafters.

The haying season usually began in mid-June, when Pa declared the crop was tall enough and mature enough—but not too mature, meaning it had flowered and grown too tall—to make a storable crop that the livestock would eat during the long, cold days of winter. Pa would hitch our team of Percherons, Frank and Charlie, to the McCormick-Deering five-foot sickle bar mower and begin slicing off the fast-growing hay crop, which consisted of a mix of legumes—alfalfa, red clover, a little sweet clover—and sometimes timothy. One small hayfield on our farm, maybe five acres, was almost all timothy, a hay crop preferred by the horses and cut and stored mostly for them. The cut hay was left on the ground to dry.

If the day was sunny and warm, and a breeze blew from the southwest, the cut hay dried quickly. It was usually ready for raking and bunching the following day—if it didn't rain and the sun wasn't hidden by clouds. In the evening on days that Pa cut hay, after the evening milking was done, we would sit on the back porch of our farmhouse and rest. If the evening was clear and still, we might hear the sound of a cowbell telling us that our neighbor's cows were out in their night pasture, as ours were.

As the mists of a cool evening rose from the hayfield, the freshly cut hay filled the air with the sweetest of smells. For farmers who appreciated the varied smells associated with farm life, the smell of hay drying made up for the hard labor required before the days of portable hay balers and forage harvesters.

The day after the hay was cut, when the morning milking was done, the cows were out to pasture, and the dew that had gathered during the night had evaporated, Pa hitched the team to the hay rake. It was called a dump rake, for it gathered the cut hay and dumped it into two- to three-foot-high ribbons that, when raking was done, stretched in rows from one side of the hayfield to the other.

Soon after the first ribbons of hay appeared, my brothers and I, each armed with a three-tine pitchfork, began the seemingly never-ending task of piling the freshly raked hay into bunches. Pa had instructed us how to build a hay bunch so it repelled water and would not flip over when the wind blew. Oftentimes if the hay had grown well, especially in the hollows that received more moisture, it was now a twisted mass of hay stems and leaves that had to be pulled apart before it could be properly piled into a Pa-approved bunch.

As the oldest, it was my job to carry a water jug to the hayfield. The stoneware water jug was dark brown on the top and tan on the sides and held about a gallon of water. It had a little handle and was stoppered with a corncob. After I had made my first hay bunch, I shoved the water jug under it to keep the water somewhat cool.

As we worked, we could stop anytime we wanted to, as Pa said, "Take a swig of water from the jug." Of course we drank directly from the jug, no cups or glasses or other such fancy things. There actually was a kind of macho way of drinking from the jug. You pulled out the corncob stopper, spilled just a little water on the ground—the theory was that the first water out of the jug was the warmest—and then lifted it up and rested it in the crook of your arm as you took a big drink. By midmorning the water was pretty warm, but as Pa reminded us, "It's still wet." Another problem, after we'd made many hay bunches, was finding the bunch with the jug under it. Every hay bunch looked pretty much like every other hay bunch to me. Pa was more discerning; as he circled the field with the hay rake, he kept an eagle eye on how good our hay bunches looked. The last thing we wanted was for Pa to stop raking hay, crawl off the rake, and say, "That's not how to do it. Must I show you again?"

The only way we boys could tell when noon was rolling around was to watch the sun in the sky; listening to the hunger pangs in our stomachs proved unreliable, as we were already starved by ten each morning. Pa carried a watch in a front pocket of his well-worn bib overalls, and it was he who decided when it was time for us to walk home for dinner. Usually by noon the field was raked, the rake driven back to the farmstead, the horses watered, unharnessed, and turned out to pasture—and then we ate. It was a huge meal, more than we ate at supper time. While we were bunching hay in the hot sun, Ma had labored over the wood-burning cookstove in a too-hot kitchen. She never complained about the heat, and Pa never did either, and we quickly learned that if we complained that it was too hot in the hayfield, our words disappeared like a cloud of hay dust on a breezy day.

In the afternoon Pa joined us with his three-tine fork to pile the hay into bunches. With him working alongside us, we were even more concerned that we construct each hay bunch as if it was the only one in the field, perfect in every way.

By supper time we were finished—twenty acres of hay bunches, each like the other, marching across the field. On our tired walk home, each of us with a three-tine fork over our shoulder, we stopped at the edge of the field and looked across its wide expanse (a twenty-acre field is eighty rods long, or more than the length of four football fields). Pa would take his fork from his shoulder, stick the tines in the ground, and lean on it. And for a moment or two, he wouldn't say anything but just look off across the field. Finally he would say, "Isn't that something to see?" For him, few things were more artistic than a hayfield dotted with hay bunches.

On day three of the haymaking process, Pa hitched Frank and Charlie to the steel-wheeled hay wagon. My brothers and I crawled on the back, and we traveled to the hayfield. Our task? To toss the bunches of hay onto the wagon and haul them to the barn, where we would store the hay in the haymow. If constructing a proper hay bunch required a certain amount of skill, constructing a proper load of hay was equally if not more important. The most embarrassing thing for the person building a load of hay was for it to tip over on the way to the barn. Such an event would lead to a story that would be told and retold whenever the neighbors gathered for threshing, silo filling, and corn shredding. No one wanted to be reminded again and again that the load of hay he built tipped over on the way to the barn.

My brothers, who were younger and thus considerably shorter than me, were responsible for driving the team from bunch to bunch and making the load, which meant arranging the hay evenly as Pa and I pitched the bunches up onto the wagon. Of course there was a knack to pitching bunches as well. It helped to have a three-tine fork with a handle longer than the one used for bunching hay. The idea was to shove the fork into about the top half of the hay bunch and then fork it up onto the wagon. This was relatively easy when there was little hay on the wagon, but much more difficult when the load was ten feet high. Not only was it challenging to lift several pounds of loose hay high over your head and steer it toward the top of the hay load, but in the process a flurry of dry hay leaves sifted down one's neck, making the process even more unpleasant.

Lighter moments sometimes broke up the monotonous work of tossing one bunch of hay after the other onto the wagon. Occasionally a snake found its way under a hay bunch, and I would discover it when I tossed half of the bunch onto the wagon. Rather than push the snake aside to allow it to crawl away, I would toss it with the rest of the bunch onto the wagon. When my brothers saw the angry, slithering reptile (usually a harmless garter snake), they did what any snake-fearing person would do—they leaped off the wagon, while I stood there laughing. Pa didn't think these episodes were the least bit funny as he stood holding the team in fear that with no one on the wagon to handle the lines, they might run away.

With the hay piled as high as we could reach, Pa and I crawled onto the wagon. He took the lines, and with the team straining against their leather harnesses, we slowly drove across the hayfield, onto the country road, and finally up the driveway on the way to the barn. But before unloading the hay, we'd stop and drink our fill of water freshly pumped from our deep well. It was cool and sweet, and nothing tasted better after working in a hot hayfield.

A couple of times during haying season, Ma would meet

us when we drove the heavily laden hay wagon into the dooryard. She had in her arms bottles of homemade root beer that she and Pa had made early in the spring. The Hires Company made a root beer extract that, when mixed with water and sugar, resulted in a fine root beer drink. Pa had an ample supply of old glass beer bottles and even a bottle capper, no doubt left over from Prohibition days when many folks made their own brews at home. What a treat it was to stop work for fifteen or twenty minutes and enjoy a bottle of cold root beer.

Pa drove the team with the load of hay onto the barn's threshing floor, where we used the mechanical hay fork, with its series of ropes and pulleys and powered by one of the horses, to lift the hay from the wagon and deposit it in the haymow. Pa insisted on working in the haymow to make sure that the hay was pushed into every nook and cranny of the mow so we could store as much hay as possible. This was called mowing back hay, and it was a hot, difficult job. If the temperature outside was eighty degrees, the temperature in the haymow, tucked under the barn roof, would be well over one hundred degrees.

My job was to set the hay fork in the hay load, repeating the move several times before the wagon was entirely unloaded. One of my brothers drove the horse on the hay fork rope—waiting to hear Pa's call of "Whoa," when the hay from the load was pulled to just the right place over the haymow. When it was perfectly positioned, I pulled on the trip rope and the hay dropped with a swoosh and a swirl of dust and hay leaves.

With the hay wagon empty, we hitched up the horses once more, backed the wagon out of the barn, and headed to the pump house. While the horses drank at the stock tank, we drank fresh water from the well. Then we headed once more to the hayfield. Pa kept his eye on the western sky, hoping that we would have all of the bunched hay in the barn before a rain shower interfered. The rule was that all hay hauled into the barn had to be absolutely dry, meaning that the hay must be allowed to dry properly, and no hay wet with morning dew or rain would be hauled in the barn. At best, damp hay would mold, spoiling it for livestock feed; at worst, it would spontaneously burst into flames and burn down the barn and everything in it. Every summer we heard a story of this happening, as farmers eager to move their hay crop into the barn failed to follow the fundamental rule. Just as we depend on fire, essential water could also cause catastrophic difficulties if not understood and reckoned with. Farmers who ignored the risk of spontaneous combustion suffered dire consequences.

# Watery Attractions

Depending on how you count them (some of the smaller ponds like the two on my farm do not make the cut), Waushara County has about one hundred lakes formed here by glaciers some ten thousand years ago. Most of them are excellent for fishing in summer and winter, ice-skating, waterskiing, and swimming, and they have long attracted people from Chicago, Milwaukee, Madison, and the Fox River Valley who come to stay in their own summer cabins or at one of the several lake resorts or campgrounds in the area.

We called these city visitors tourists—they were easy to spot because many of them wore short pants, and no farmer would be caught dead wearing short pants. My brothers and I also thought some of the city folks talked funny. We, of course, thought we knew how to speak properly and found people who had accents different from ours a bit unusual. But our parents knew that these lake people spent a bunch of money in the area, buying the eggs and strawberries that Ma delivered to the Mercantile in Wild Rose, eating at the little restaurants in town, buying gas at the gas stations, drinking beer in the taverns, and buying ice cream and candy before taking their seats near us at the Tuesday night free shows in Wild Rose. We were taught to not make fun of these city visitors who sometimes talked funny, dressed even funnier, and had too much time on their hands. In fact, Pa had my brothers and me convinced that we should feel sorry for them. He told us it was too bad they didn't have enough work all summer to keep them busy.

For our family, the idea of a vacation was unheard of, no matter if it was summer or winter. The cows came first, and they required milking twice a day, 365 days of the year. We did take advantage of the rare times when wet fields prevented us from haying to fit in a few hours of fishing.

And once each summer, on the Fourth of July, our parents allowed us to spend a whole afternoon at the lake, swimming and splashing and picnicking along the shore. Early July was midway into the haying season, so we spent the morning working in the fields. Even though Pa agreed that celebrating the Fourth was important, we could devote only half a day to such activity, for the barn had to be filled with hay—as farmers said in those days, "Make hay when the sun shines," and usually the Fourth of July was a sunny day.

But at noon Pa called a halt to haymaking, unharnessed the horses and let them out to pasture, went into the house to clean up and change his overalls to a clean pair, and instructed my brothers and me to do the same. Ma had been busy that morning preparing a picnic lunch of German potato salad, one of her specialties, plus cheese sandwiches on homemade bread, baked beans using beans we had grown the previous season, dill pickles from the shelf in the cellar, Jell-O, lemonade made with real lemons, and a strawberry pie with berries from the tail end of the strawberry season.

We piled into our black 1936 Plymouth and headed to Silver Lake. There we found a picnic table under the huge pine trees. Then, while Ma was laying out the lunch and Pa was relaxing in the shade, my brothers and I were off to the bathhouse, where we could change into our swimsuits and store our clothes in a little basket for ten cents each. (Pa thought the cost was outrageous, as he considered swimming a right that we shouldn't have to pay for.) All afternoon we swam in the lake, crawling up the water slides and sliding down over and over. We didn't venture as far as the swimming raft anchored in the lake, since we couldn't swim that well.

In late afternoon we returned home and shot off a few firecrackers. Firecrackers were legal in those days, and somehow we scraped together enough money to buy some "lady fingers," as they were called, little but loud firecrackers that were tied together so when we lit one the whole array went off, one after the other, like the sound of an automatic rifle.

When the milking was done and the cows were turned out to pasture, we gathered on the seldom-used front porch to catch glimpses of the fireworks at Silver Lake, some six miles from our farm. It had been a day we would fondly remember, and we would look forward all year to doing it again.

We also hosted city visitors a few times each summer. Both my mother's relatives from Wisconsin Rapids and family friends from Chicago came to spend time at our farm. On a hot Sunday afternoon, when we could get away from our farm chores, we liked to take our visitors to the nearby fish hatchery. In 1907 the state of Wisconsin had purchased property in Wild Rose from a Mr. G. Jones for $2,500 for the purpose of establishing a fish hatchery. The site was perfect for such an endeavor, as natural springs flowed out of the hills to the west and eventually flowed into the Pine River below the millpond dam at Wild Rose. In a series of concrete raceways that were about six feet wide and a few feet deep, we could view young trout, from the tiniest fingerlings to those a foot or more long. The hatchery fish eventually found their way into the lakes and streams in this part of Wisconsin, ensuring that fishermen who came from near and far would land a fish or two for their efforts.

What I marveled at most in those vast raceways were the gigantic lake sturgeons swimming among the trout. Each raceway had one or more of these ancient fish. Covered with skin instead of scales, with a spadelike snout, the sturgeon plied the bottom of the fish hatchery raceways, keeping things clean. When a hatchery worker came to toss feed on the water, the rambunctious trout exploded to the surface in quest of a meal. The sturgeon remained on the bottom, content to live and feed there. Some of them were four or five or even six feet long, and I was told some weighed more than two hundred pounds—giants among the tiny trout with whom they shared a watery home.

Of course fishing from the hatchery was not allowed, as the fish were intended to stock nearby waters for fishermen. Yet every summer I read in the *Waushara Argus* about some person getting caught fishing in the fish hatchery. It was easy to catch fish there—and, as it turned out, easy to catch the fishermen as well.

# Summer Thunderstorm

By the summer of 1941, drought had returned to our part of the country. That August we had no rain to speak of, except for a brief shower or two. Our sandy farm began showing the effects of the dry spell. The hilltops in the cow pasture were brown with dead and dying grass. The only green pasture was in the hollows, and that was disappearing. The lower leaves on the corn growing on the hillsides were turning brown and dying. The potato plants that had looked so promising in early summer were now withering and curling. The garden crops were suffering similarly. Ma became concerned, for it was our garden that provided most of the vegetables our family ate throughout the year.

Thankfully, with the gasoline pump engine pumping water from our deep well, the livestock had plenty of drinking water, and we had sufficient water for family use. But without rain, the possibility of a poor corn crop loomed, which meant less corn in the corn cribs. If the pasture grass completely dried up, we would be forced to feed the cattle hay that we had stored for the winter. If our twenty-acre potato crop—our main cash crop—failed, we'd lose the source of a little extra farm income. When it didn't rain everything suffered, the farm animals, the crops, the wild animals and birds, and the people. Pa's mood turned sour, and Ma grew quiet.

Every evening, after the cows were milked and turned out to the night pasture, Pa and I would stand by the barn, facing west. Pa studied the western sky, hoping for some hint that rain might be coming our way. He examined the cloud formations. He kept track of the wind direction. And as the sun set each evening, a red ball of misery, Pa would shake his head and say nothing, which I knew meant we were facing another day with no rain.

Finally one evening, Pa had a more hopeful look on his face.

"See that bank of black clouds on the horizon?" Pa said. I looked, but all I could see was a blazing hot sun that had begun its descent below the horizon.

"Maybe tonight. Maybe a shower tonight."

As evening turned into night, Ma lit the kerosene lamp that sat at the center of the kitchen table. She would blow it out an hour or so later.

"You boys better get up to bed," she said. My brothers and I climbed the stairs to our bedrooms, me to the little bedroom on the end of the long upstairs hall, the one in the corner with one window facing south and the other east, my brothers to the big center bedroom that had double windows facing east.

Both windows in my bedroom were open, but there was

not a hint of a breeze and the room was stifling hot. I piled my clothes on a chair and was almost immediately fast asleep.

I awakened to the sound of distant thunder. Crawling out of bed, I glanced toward the southwest and could see flashes of lightning tearing across the ever-blackening sky. Pa was right, rain was on the way. But now I feared what I was sure Pa also feared: that the storm would bring wind, wind strong enough to topple the corn plants, tear the wooden shingles from the barn roof, or, even worse, topple trees or blow over an outbuilding. The barn was mostly full of hay, so it would take a substantial wind to move it. Another fear was hail—chunks of ice, sometimes as big as baseballs, that would shatter the corn plants, flatten the potato crop, and ruin the vegetable garden.

As I watched from my bedroom window, I hoped for rain that we so desperately needed but worried about the storm's destructive power. And I wondered why the sky had to be so angry to drop some rain on our droughty farm community.

The storm crept closer. Flashes of lightning nearly overlapped each other, and the booming thunderclaps became ever louder. I heard my father's voice: "Boys, wake up. Wake up and come downstairs. Big storm coming."

I could remember only two or three other times that Pa had awakened us during a thunderstorm. He offered no explanation, but I knew why. Perhaps even more than drought, he feared fire. If lightning should strike the house, or even worse, the barn, he wanted us all to be awake.

I pulled on my bib overalls, stopped by my brothers' bedroom to make sure they were awake and up, and hurried down the hallway, which the lightning flashes made as bright as day. An enormous clap of thunder rattled the windows in the hallway as I came to the top of the steep stairway. I found Ma waiting in the dining room. "Did you close your bedroom windows?" she asked me. I didn't answer but immediately ran back upstairs to do that. Pa was in the kitchen, looking out the window facing south, toward the barn. Another flash and another roar of thunder, now coming at almost the same time, and then the rain began falling in a torrent. A real "gulley washer," as our neighbor Bill Miller would later say.

And then the wind came with a rush—not a wind bent on causing damage, but a wind that cooled and soothed and made a hot summer night bearable.

I watched out the kitchen window with Pa. He was smiling. The angry part of the storm, the dangerous lightning and loud thunder and the possibility of hail and wind, had moved to the east. We could still see it, but soon we could no longer hear it. The rain continued. A steady rain. Without exaggeration, a lifesaving rain.

"You boys can go back to bed now," Pa said. "It's three o'clock; you've got a couple more hours of sleep before it's time to get up."

Back in bed I listened to raindrops splattering against my bedroom window, and I felt good. All the living creatures on our farm felt good after a soaking rain. For Pa and Ma, a rain such as this one changed their mood from despair to hope. Calling it joy would be going too far, for the best a farmer could feel during the years when I was growing up was hope. Hope was in short supply during the Depression years and World War II. But a good rainstorm could provide a substantial amount of it.

# Fire at the Farm

Mrs. McKinley Jenks, who lived a quarter mile from the Chain O' Lake School, came running over with the news. It was a warm September afternoon in 1946. Mrs. Jenks burst into the schoolroom and interrupted Miss Thompson, who was in the middle of a reading lesson with some of the upper-grade students. I was in seventh grade and a part of the group. My twin brothers, Donald and Darrel, were in third grade.

Miss Thompson quickly took me aside and shared what she had just heard. "Mrs. Jenks just heard on the party line that your barn is on fire."

It was the worst possible news. Like all farmers in our community, we milked our small herd of dairy cows in the barn and housed them there during the cold months of winter. We also stored hay for the cows there, in the hay-mow. Occasionally when a farmer's barn burned, the cattle housed inside perished as well; because it was September this was unlikely, as our cows were out on pasture, some distance from the flaming barn. The loss of a barn was horrendous, sometimes so disastrous that a farmer was forced to move to a different farm.

In 1946 the Wild Rose Fire Department was not yet allowed to fight fires outside of the village, so when fire struck on a farm, it had to be fought with the help of the neighbors, who never had enough water for the task. The neighbors hauled water from their stock tanks and pumped from their wells and filled ten-gallon milk cans with the precious water, which they then hauled to the farm on the backs of pickup trucks, on wagons pulled by John Deere and Farmall tractors, and even on rickety trailers pulled behind old 1930s automobiles.

Usually these neighborhood firefighters failed to save the barn. The best they could do was to save the farmhouse and other nearby buildings—and too often they failed at that as well, especially if there was a substantial wind that blew live embers on the wooden roofs of these nearby buildings.

Upon hearing the news, I immediately wanted to grab up my twin brothers and run home to help with the firefighting. But Miss Thompson said no. "Firefighting is dangerous—it's best if you and your brothers stay here until school ends this afternoon."

All I could do the rest of the afternoon was stare at the old Regulator clock on the wall and wonder what I would see when I returned home that afternoon. I expected to find the barn reduced to a pile of smoldering ashes. I hoped that Pa and the neighbors were able to get the calves and the herd bull out of the barn before the flames consumed it.

Four o'clock finally arrived, and my brothers and I ran

all the way home. As we approached the farmstead, we could see a cloud of gray smoke rising from where we knew our barn had stood. We expected the worst. But as we got closer, beyond the cloud of smoke I could see the barn. It was still standing, and it did not appear to be on fire. Something was burning fiercely in the barnyard, however, just a few hundred yards from the barn.

I recognized several of our neighbors—Guy York, Alan Davis, Bill Miller, Frank and George Kolka, Freddy Rapp, the Macijeski boys, Arlin Handrich, and others I didn't know. They were tossing pails of water onto the nearby pump house, wetting down the roof and walls closest to the roaring flames.

At first I couldn't tell what was burning. It obviously wasn't the barn or another of our farm buildings. Then it occurred to me that I couldn't see the huge straw stack that was piled a few feet northeast of the barn. Sure enough, it was the straw stack that was burning furiously in the middle of the barnyard, with no one making any attempt to put it out. The men knew that pouring water on the straw pile might put out the flames, but the water would spoil the straw.

But how had the huge straw stack moved from only a few steps away from the barn to the middle of the barnyard?

Later I learned what happened. When the general ring had gone out to all the farmers in the community and the neighbors began arriving, they immediately saw smoke coming from the straw stack. It was not yet fully engulfed in flames. John Macijeski suggested pulling the straw stack to a location where it could burn itself out without igniting the nearby barn or any other buildings. But the question was how to do it, for the straw stack was probably twenty feet high and many feet around.

John Macijeski spotted a length of quarter-inch steel cable that we had previously used as part of a system to move cow manure from the barn to a manure pile in the barnyard. John suggested to the men that they wrap the steel cable around the smoldering straw stack and pull it away from the barn. Most of the men didn't think the idea would work, but they didn't know what else to do. With the cable completely encircling the straw stack, Pa hitched his Farmall H tractor to where the cable ends came together, and Bill Miller hitched his John Deere B in front of Pa's Farmall. With a little wheel spinning and both tractors roaring, the big straw stack began moving. When they had it positioned in the middle of the barnyard, a goodly distance from any building, Pa and Bill quickly unhitched their tractors and drove them out of the way. A couple of other men pulled the cable away—and then, as moving the straw stack had fed the smoldering fire more air, it burst into flame, with a cloud of gray smoke rising a hundred feet into the sky and drifting eastward on a slight western breeze. Now the men worked quickly to wet down the pump house, which was closest to the burning straw stack.

The straw stack was mostly burned to the ground by sunset, as the tired and dirty firefighters drifted away toward home to do their evening chores. Pa went around and shook each man's hand, thanking them for saving his barn.

"You would do the same for us," each man said in response.

Later we figured out what had likely caused the fire. That year we did not have enough space in the barn to house the young stock (cattle that were not old enough to add to the milking line in the barn), so Pa had constructed a rustic shelter of old boards, a few cedar fence posts, and unused woven wire fence. He would cover it with the straw stack to provide the animals insulation and shelter from the cold winter wind.

But Pa had made a mistake. Before building the straw stack, he had covered the shelter with alfalfa hay that had not been sufficiently dried. Later, after threshing, he covered it with dry straw. It was the alfalfa hay that had spontaneously burst into flame and caused the entire straw stack and the shelter beneath it to burn. No young stock had been in the shelter at the time of the fire because by this time they were out to pasture with the milk cows.

Had it not been for the help of the neighbors, who hauled water tirelessly and came up with the innovative idea to haul the straw stack away from the barn, we would likely have lost our dairy barn and the pump house as well. Maybe even the farmhouse. It was a frightening day, but one that could have ended much worse.

# First Fall Rain

Pa said he could feel it in the air. It was a rather chilly late September afternoon, and the grain was threshed, the oat bins running over with fresh grain, and the straw stack piled high. The silo was filled and the barn loft stacked with hay. The remaining corn had been left to ripen, and after that we'd have to do the fall plowing. We were mostly on schedule, if a farmer's year can ever be on schedule—because something usually came along to turn well-thought-out plans on end.

And now Pa sensed the first fall rain on its way. The first hard rain of fall was always welcome, as many summers were dry on our sandy farm. Fall plowing would be much easier with well-soaked ground. And the alfalfa and clover that Pa planted with the oats as a nurse crop were just sputtering along as they waited for rain. A good soak would give new life to these fragile legumes before the long winter froze the ground from November to late March.

We were mending fence that chilly late September afternoon, a task we always did between other tasks deemed more important. Of course, sometimes fence mending became an emergency task, especially if our herd of cows broke through the fence and ended up in the neighbor's cornfield.

But on this cloudy Saturday, we were taking our time, setting new posts and stringing new barbed wire along a stretch of fence where the posts were rotting and the wire was old and rusty. Fence making was never easy. Ninety-nine percent of the time when digging a new posthole, you struck a rock. Then you either figured out a way to remove the rock or you moved the hole slightly in one direction or the other to avoid the rock. It didn't help to complain about the rocks, which I often did. Pa reminded me that rocks were a part of a farmer's life and I'd best learn to live with them. I never got quite that far in my relationship with these gifts left by the glacier some ten thousand years ago.

As we worked, I thought how it was always easier making fence on a cool day like this one compared to a ninety-degree midsummer day. Pa stopped occasionally and glanced to the west, to the darkening cloud bank. "Yup, I believe it's gonna rain," he said.

I was thinking, the sooner the better, for if it began raining we would gather up our fence-making tools, toss them on the wagon, and head the team of horses and the wagon toward home.

Less than a half-hour later Pa proved right in his prognostication, as I felt the first drops of a cold rain strike me in the face. "Best load up and go home," he said. On our way home, the rain became heavier. I pulled my cap down and pulled up the collar on my denim jacket, which was far from waterproof. There was no thunder, no lightning, just an ever-steadier, bone-chilling, soaking rain.

Pa was smiling as he drove the team down the crooked lane that led from the far reaches of the fields back to our barn. Pa always smiled when it rained, for he knew better than I did how important the rain was, no matter what season of the year. All I had on my mind was a dry, warm place.

Arriving at the farmstead, Pa unhitched the team, led them into their stalls in the barn, and unharnessed them. Rainwater dripped from their brown coats, but they seemed not the least bit bothered. They had had an easy afternoon, toting the wagon with a light load to the far field, standing around most of the afternoon, and pulling the wagon back. Nothing at all compared to plowing with a one-bottom walking plow, which would be their task in a couple of weeks.

"Hafta keep the cows in the barn tonight," Pa said after taking care of the team. This would be the first time since way back in April that the cows would stay overnight in the barn, marking the changing of the seasons better than any calendar might do.

Keeping the cows in overnight meant carrying in forkful after forkful of fresh straw from the straw stack and spreading it all across the barn floor where the cows would spend the night confined in their stanchions. As we worked, I noticed that the cows had returned to the barn from the back pasture on their own. They stood huddled outside the barn door, cold rainwater dripping from their black and white coats.

With the straw in place, not only did the barn look different, but it smelled different as well. The fresh oat straw had a subtle, rich aroma that I liked. It contrasted with the rather medicinal smell of the lime that we spread in the walkway behind the cows to keep the barn smelling fresh.

Once we let the cows into the barn and they found their way to their stanchions—each cow had its own place in the barn and knew to find it without any encouragement—the smells in the barn changed again. Now the fresh, earthy smell of oat straw fought with the rather harsh smell of wet cowhide.

With the cows settled in, we walked to the house for supper, hoping that the cows would be dry enough later that we wouldn't feel rainwater dripping down our necks as we sat under them with milk pails captured between our knees.

After supper Pa and I returned to the barn. With the cows milked—they had mostly dried off by this time—I crawled up the ladder to the haymow and tossed loose hay down the hay chutes to the cattle below. This would be their first feeding of dry hay since last spring and would mark the beginning of feeding them hay throughout the long winter months.

The following morning, with the rain stopped and the clouds moved off to the east, we turned the cattle out to pasture once more. Perhaps they could stay out overnight tonight, perhaps not, depending on how far the temperature would fall. If Pa suggested the cows should stay in the warm barn again, we were ready. The straw was in place, and I once again had gotten the hang of throwing down hay from the haymow.

The first fall rain had left us about an inch—marked by how much was in the dog's dish that always sat outside next to the kitchen door. Farm people like my family knew we needed the rain to sustain our overwintering hay crop and to make fall plowing easier. For my part, I both dreaded its arrival, for it marked the end of summer, and looked forward to it, for autumn was my favorite time of year.

# Roche A Cri River

I could hardly wait for hunting season of 1946 to roll around. Although I had been hunting squirrels and rabbits since I was ten years old, since turning twelve in July I could now do it legally because I could buy a hunting license. Even more important, I could now accompany my dad and our neighbor, Bill Miller, on their annual trip to nearby Adams County for deer hunting. At that time there were just a handful of deer in Waushara County but many more in Adams County, the next county to the west.

On a cool, cloudy November day, opening day of the 1946 deer season, we got up at four-thirty to milk the cows and do the assorted barn chores. Then we ate a big breakfast, donned our red hunting gear—I had a new red hunting coat—and headed off to the place in Adams County where Pa and Bill had hunted for many years. I was to use my father's 12-gauge, double-barreled shotgun, which weighed a ton and kicked like a horse. I had used it for rabbit hunting so I knew its characteristics. Pa and Bill had

30–30 deer rifles, lighter and more effective than a double-barreled shotgun, but I didn't complain. Not a bit. With my new hunting license in my pocket, I was off on my first deer hunting trip and feeling elated about it all.

Upon arriving in the township of Rome in Adams County, the community where my father was born and had attended a one-room log school, we found an open field near my dad's home place, parked the car, and loaded our guns, and the deer hunt was on.

Pa and Bill said they would walk west to a spot about a mile away. They told me to wait for a few minutes before I started walking in the same direction. I understood that I, the newcomer to deer hunting, was to walk through the woods and scare out any deer that might be there so Pa and Bill could bag one.

As we stood on the banks of the Roche A Cri River, a fine, slow-moving stream that threaded through mostly abandoned farmland and woods, Pa said, "Follow the Roche A Cri, and you won't get lost."

It hadn't occurred to me that I might get lost, but I could quickly see that it was a clear possibility. Few people lived here, the dusty dirt roads were sometimes more than a mile apart, and the woods were dense and thick and new to me.

Pa and Bill disappeared on their way to their chosen place a mile away. I walked along the river's edge for a few minutes, lugging the 12-gauge, which got heavier as I walked. I leaned the shotgun against a big pine and sat down on the banks of the river, resting and watching the water, which was moving slowly, more slowly than I remembered about most rivers and streams.

The day was cloudy and rather dreary, but not cold as some late November days can be. I dozed off for a bit, somewhat mesmerized by the river and already tired from carrying the heavy shotgun. Then a sharp report, not too different from the sound of a rifle, came from the river. Instantly I was awake. Was someone shooting? Should I gather up the 12-gauge and be at the ready as a deer might come bounding out of the gloom?

Before I moved, I saw the source of the loud sound. It was not a rifle, but a big beaver slapping its tail on the water to warn its beaver family that it had spotted an intruder on the riverbank. I didn't move, and the big beaver swam by me, its head out of the water, the rest of its body submerged. Glancing upriver, I spotted the dam the beavers were building across the Roche A Cri—and now I knew why the river water was moving slowly.

For more than a half-hour I sat watching the beavers work, carrying sticks in their mouths and depositing them at the dam. I was fascinated with how hard they worked and remembered the old saying "busy as a beaver." They were indeed busy.

Finally deciding that Pa would wonder why I was taking so long to cover the mile or so of river frontage, I began walking. I spotted Pa up ahead.

"Did you get lost?" he asked.

"Nope, just took my time," I answered.

When we were back home I told Pa about watching the

beavers working in the river. I thanked him for taking me along deer hunting. Although I hadn't shot at a deer—hadn't even seen one—watching the beavers hard at work in the Roche A Cri made up for it.

# Ice Fishing

Ice is an integral part of winter, especially in the upper Midwest. It is both cursed and praised, often on the same day. Water freezes in pipes and ruptures them. Ice-covered roads make driving treacherous. And who hasn't slipped and fallen on an ice-covered walkway? But on the positive side, ice-skating is great fun, and so is ice fishing.

Pa wouldn't go ice fishing until the ice was at least a couple inches thick. He was not one to defy nature and risk falling through the ice, with a real chance of drowning or dying from hypothermia before another fisherman was able to rescue him. I knew it was time to commence ice fishing when Pa dragged out his tip-ups, handmade wooden devices that he set over an open hole in the ice with a heavy green fishing line attached. We fished for northern pike, heavy buggers that can weigh up to twenty pounds, and when a fish grabbed the minnow that hung on the hook on the end of that green line, the homemade device tipped up, revealing a red flag. This told us a fish was on the line—or at least had yanked hard on the bait.

Our favorite ice fishing lake was Mt. Morris Lake, really a large millpond that had once provided waterpower for the Mt. Morris Mill, which still stood near the dam at the end of the lake and still ground cow feed for the nearby farmers when I was a kid.

On a cold Saturday morning in January after all the barn chores were done, we headed for the lake. Pa parked the old Plymouth on the roadside and we walked the remaining half-mile carrying the tip-ups in a gunny bag, one that had holes in it and no longer served to carry cob corn or oats to the mill for grinding. Two tip-ups for Pa and two for me; when my brothers were older we stuffed eight tip-ups in the old gunny bag. Over his shoulder Pa carried an ice chisel that Arnold Christensen, the blacksmith in Wild Rose, had fashioned from the rear axle of a Model T Ford car. I carried the minnow bucket that contained the couple dozen minnows we had bought from the bait shop in Wild Rose and a lunch bucket with sandwiches, cookies, and coffee.

Pa had fished Mt. Morris Lake many times, more often in winter than in summer, so he had a good idea where the northerns might be lurking. With the ice chisel he set to punching holes about eight inches across through the lake ice. Then he set up our tip-ups, one after the other, in a straight row across one corner of the lake.

With the tip-ups in place, we found a sheltered corner in the woods on the shore, out of the northwest wind that made a moderately cold day uncomfortably cold out on the ice. I helped Pa gather some dry firewood and a small armful of dead cattails for our campfire. Soon the little fire was blazing, sending up a trickle of sweet-smelling smoke that drifted across the lake where our tip-ups waited. We huddled around the fire, keeping warm and enjoying the day. Fish in cold water are rather lethargic; they move slowly and they aren't near as hungry as they are in summer. Thus the tip-ups would remain inactive for long periods, offering more time for sitting around the campfire, looking off across the lake, and enjoying a winter day.

By eleven o'clock other ice fishermen arrived, neighbors and friends whom we worked with on threshing crews, at silo-filling bees, and at other farm activities requiring extra help. Jim and Dave Kolka wandered over to our campfire and found seats on the end of a log. My Uncle Wilbur found another seat.

As we added more wood to the fire and occasionally glanced at our dormant tip-ups, the stories began flying. Uncle Wilbur, known for embellishing a story a bit more than most, began sharing one of his tales—the one about the time he started to pull a twenty-pound northern pike from these waters but needed the help of three men with ice chisels to enlarge the hole. Wilbur waited patiently for the work to be done, with the big fish tugging and pulling and trying to free itself. Finally the fish hole was large enough, and Uncle Wilbur hauled the big fish up onto the ice.

"Just like a pig on ice, it was," related Uncle Wilbur.

Everyone laughed, knowing that the fish probably weighed no more than ten pounds (still a good-sized fish) and that it likely took only one person to enlarge the hole.

At noon, we opened our lunch buckets and toasted our

cheese sandwiches over the campfire. Nothing could be better than toasted homemade bread, spread thick with butter and a slice of cheddar cheese and tasting faintly of oak smoke.

So far we'd had but one tip-up flag go up, and that was a miss. Jim Kolka, always a little braver than the rest of us, announced that he wanted to go exploring to see how thick the ice was where a little stream poured into the lake. We could see the open water where the warmer stream water entered the lake.

Pa said, "Jim, you watch out, the ice is pretty thin near that open water."

Jim set out to prove that Pa was wrong, walking confidently toward the small pool of water that was black against the snow-covered ice. All eyes now turned toward Jim. Storytelling ceased. The only sound was that of the snapping and cracking of the little campfire.

Closer and closer Jim walked toward the open water. He turned back occasionally to make sure his audience around the campfire was watching. We were, of course.

Then there was a crack and a splash and Jim fell through the ice. Because he was close to shore, the water was only about chest deep—so when the splashing ceased we saw Jim, with a very surprised look on his face, standing on the bottom of the lake with his arms resting on the edges of the hole through which he had fallen.

"Get me outta here," he yelled as he discovered he could not lift himself out of the hole. The freezing lake water had almost immediately soaked through all of his clothing, and he was not only humiliated but suffering the ice-cold consequences.

Pa immediately grabbed the hatchet we had brought along to cut campfire wood and walked a few yards along the shore to where some tall aspen grew. He quickly chopped down an aspen that was three inches or so in diameter and about twenty feet tall.

Pa yelled to Jim to quit thrashing around, as he was only making things worse by breaking more ice around the hole. "Stand still," Pa yelled. And Jim did.

A couple of us walked toward Jim, pulling the cut aspen behind us. Pa walked in front of us, carrying the ice chisel and chopping into the ice every few steps to determine its thickness. When the ice was about two inches thick, Pa told us to stop, get down on our stomachs, and push the tree toward Jim.

Jim grabbed the end of the aspen, and with some effort we pulled him up onto the ice. We hustled him toward the campfire, to which the others had added more firewood. We helped Jim take off his clothes right down to his long underwear, and we sat him down by the fire. His teeth were chattering. After an hour or so, when his clothes were reasonably dry, Jim put them back on and got back to fishing.

Of course, we never let Jim forget that day. No matter if we were together for threshing, filling silo, or shredding corn, someone always asked, "Say, Jim, can you tell us about the time when you tried to walk on water?"

We laughed about it, but all of us who witnessed it that cold January day recalled how unforgiving water can be.

# Spring Breakup

After a long, cold winter, we looked forward to those first warm days in March, when the snow became mushy and little trickles of meltwater oozed out from the bottom of the snow piles, indicating that spring was on its way.

The twenty-acre field at the northeast corner of our farm had a small gully cut into the side of one of its several hills. Into this gully we dumped fieldstones that we picked from the field over the years. Pa's hope was that the stones would prevent or at least slow down soil erosion in this spot.

The gully was one of my favorite places to visit on a warm late-March day when the snow was melting. As the meltwater ran over and around the stones, it made the most wonderful music, a subtle tinkling. It was the music of spring, a sound of joy, as if the entire landscape was celebrating the end of winter.

Spring breakup presented challenges as well. In 1949 we'd had a very snowy winter, and in mid-March the temperature soared into the seventies for a couple days, rapidly melting the snow. And then it rained. The frost had not yet gone out of the ground, so the meltwater and rainwater accumulated in all the low places, flooding several of the well-traveled roads. On one section of County Highway A, our main route to Wild Rose, the water was three feet deep—too deep for most cars, but not so deep that it prevented the milk truck from making its daily rounds to pick up farmers' milk. The school bus that toted me to the high school in Wild Rose couldn't make it through, though the high school remained open. I didn't want to miss school, so I walked the four and a half miles to Wild Rose, often along flooded places in the road. I had to get up a bit earlier in the morning to make sure my barn chores were done before heading off for about an hour's walk to school.

After several days of warm, rainy weather, winter returned, coating all the newly formed ponds with ice and making road travel even more difficult. My Uncle Wilbur was certain his new Ford milk truck had the power to travel through these ice-coated ponds. When he drove onto the ice-covered County A flooding, the ice broke and his truck settled to the bottom of the pond, smashing the grille.

Warm temperatures finally returned, and the ice melted, the ponds blocking the roadways disappeared, the frost went out of the ground, and the school buses once more made their daily rounds.

Water, whether in the form of snow, ice, or liquid, is not always predictable. And thus spring in the country is never boring. Feelings of joy and despair can be mighty close together during a country spring.

# Lakeside Lodge

I couldn't wait to turn sixteen and get my driver's license—
an important milestone, even though like most country kids
I had been driving tractors and pickup trucks since I was
twelve years old. Now I could drive on the country roads
and even the state roads without fear of having the lone
traffic cop in Waushara County stop me and ask to see a
nonexistent driver's license.

An even more important milestone was my eighteenth
birthday, for at that age I could legally drink beer. Perhaps
most important of all, at age eighteen I could enter the
dance hall at Lakeside Lodge, located on the south shore of
Fish Lake near Hancock, Wisconsin. Lakeside Lodge had a
beer bar, and there we eighteen- to twenty-year-olds could
congregate with our dates and friends, drink beer, and talk
smart as we tried to impress our gals with our worldliness.

Lakeside Lodge held a dance every Saturday night, year-
round, with the possible exception of December 24. Tickets
cost fifty cents. The live orchestra—always a polka band,
for this was German and Polish country—began arranging

themselves on the stage on the far end of the hall around eight-thirty, as the dance began promptly at nine o'clock.

Many of western Waushara County's hardworking farmers and their wives went to Lakeside Lodge on Saturday night to dance, drink a little, and celebrate the passing of another week. Of course, not all farmers liked to polka dance or even knew how. My pa was one of those who had no interest in dancing, often wondering aloud why so many people would choose to "make fools of themselves prancing around a dance floor."

But there on the shores of Fish Lake on a hot summer evening, young and old gathered to hear a polka band play the wonderful music associated with their heritage: polkas, old-time waltzes, schottisches, circle two-steps, and an occasional "drag along" tune, as some referred to those songs played for folks who didn't know how to dance but wanted an excuse to snuggle a little with their dates on the dance floor—and show off that they had brought the prettiest girl to the dance.

Beer was sold in little seven-ounce bottles, five for a dollar. But for those of us fortunate enough to have dates, dancing was more important than beer drinking. Even better was walking on the sandy beach with your date on a moonlit night with the lights of the dance hall reflecting on the still waters and the sound of polka music echoing across the lake. What could be more romantic? Especially if the band was playing a tune such as "The Tennessee Waltz": *I was dancing with my darlin' to the Tennessee Waltz, when an old friend I happened to see...*" Or the "Blue Skirt Waltz":

*"I wandered alone one night, till I heard an orchestra play. I met you where lights were bright and people were carefree and gay..."* The sights and sounds reflected and echoed across the dark waters of Fish Lake, a place where romance could develop, progress, and become interesting (words my buddies used in asking how things had gone on the beach).

Then, almost predictably, the mood changed as the band began playing "The Beer Barrel Polka": *"Roll out the barrel, we'll have a barrel of fun! Roll out the barrel; we've got the blues on the run!"* Or perhaps what my buddies and I called the "Butcher Polka": *"Put your arms around me, honey, hold me tight."*

With the romantic mood broken, the young couple returned to the dance hall and resumed dancing. But neither forgot that special moment on the sandy shores of Fish Lake with the sounds of wonderful music blending with the sounds of a late summer evening, the crickets and the bullfrogs and the gentle lapping of water on the sand, and the sight of a big yellow moon hanging low over the lake.

# Tent Camping

From the time I was a little kid and began reading books like Horace Kephart's *Camping and Woodcraft*, I wanted to go camping. I wanted to sleep outdoors in a tent and listen to a gentle rain on the canvas roof. I wanted to hear the wild animals and birds of the night and feel the coolness of the evening after the sun had set and an evening breeze washed over the land.

Somewhere I'd gotten hold of a Boy Scout manual with its assorted descriptions of camping activities. But there was no Boy Scout troop within miles of our farm. So I saved my hard-earned money from picking cucumbers and green beans that we raised as cash crops, and I ordered a tent from the Sears, Roebuck catalog. For days I waited for the tent to arrive. I even picked out the spot where I would pitch my new tent, a place near the big woods that was only a couple hundred yards from our farmhouse. Instead, I found a letter in our mailbox stating that the tent was unavailable and my money was returned. My hopes for sleeping in a tent were shattered.

Finally, at the ripe old age of nineteen, I found myself camping—bivouacking—at army Camp A. P. Hill in northern Virginia, where I was sent for basic training. For the first time in my life, I would be sleeping in a tent, a pup tent, which I would share with a fellow soldier from New York City. It was 1954, and we were using leftover equipment from the Korean War and even some from World War II.

A pup tent in those days was anything but fancy. Each soldier carried half of a tent in his backpack. When evening rolled around, two buddies buttoned the halves together, and they had a tent. It was a bit cozy for two, but it worked. I loved it. My tent mate from New York City, on the other hand, hated the tent and detested the out-of-doors. It was the first time he had been outdoors at night, away from city lights, and he was scared to death. He quizzed me about wild animals that might attack us during the night. He asked whether snakes would slither into our tent while we slept or if a strong wind might come up and blow down our canvas shelter.

As my partner fretted, I used my army-issue foldup shovel to begin making a trench in the red clay soil all around our tent. He asked me what I was doing, and whether it was a strategy to keep snakes out.

I laughed. "It's to keep the rain out," I said.

"What rain?"

"You'll see," I said. "It feels like rain."

Now he laughed. "Feels like rain, huh?" From the look on his face I could see he thought I was some kind of country nut.

It was several hours later when I heard the first growl of thunder. During a flash of lightning I glanced at my watch—it was midnight. My tent mate was fast asleep, no doubt dreaming of snakes crawling under his olive drab wool blanket. Then I heard the first raindrops, like someone beginning to beat on a drum, slow at first, then faster and faster, until the drumbeat was a crescendo of falling rain. To me it was a beautiful, primitive, wonderful sound, sending me back to our haymow on a rainy day. But this sound was much louder, closer, more intimate.

My tent mate awoke with a start and sat up.

"What's that noise?"

"It's rain," I said. "Go back to sleep."

"What's gonna happen to us?"

"At worst we might get wet, but if the trench I made around the tent works, we'll be fine." And we were.

It was the first of many times when many of my fellow soldiers grumbled and complained about rain or sleeping outdoors. Yet I felt just the opposite. I wanted to spend more nights in the rain in a tent. The US military granted my wish, as I spent ten years in the army, the majority of that time in the reserves but some on active duty and at two-week summer camps at bases around the country. While my soldier friends almost always complained about rainy nights spent in tents, I always enjoyed the raindrops drumming on the canvas making music as powerful and beautiful as that made by the finest symphony orchestra.

After Ruth and I married in 1961, I wanted to introduce my bride to tent camping. It would be a new experience for her, as she had never spent the night in a tent. We lived in Green Bay at the time, and I watched the *Green Bay Press Gazette* for ads offering used tents. Finally I saw one: "Umbrella Tent. One center pole. Good condition. $40.00."

Ruth and I found the address and soon were looking at the tent, which had been set up in the seller's backyard. It had a heavy wooden pole at its center, with a series of guy ropes leading out of it, and stakes pounded in the ground all around it. It even had a canvas floor, something my army tents did not have.

"Would you take $37.50 for the tent?" I asked.

After a little hesitation, the seller said yes, and we had a new canvas tent that, when folded up, took up half the space in the trunk of our car. I made reservations for two nights later that summer at a campground on the east end of Lake George, not far from Rhinelander.

When Ruth and I arrived at the campground late on a Friday afternoon that July, we saw that the campground spots were nearly all taken, but one had been reserved for us. Ruth and I set up our tent, put the center pole in place, tied all the guy ropes, and pounded in the stakes all around. Then I looked off to the west and saw a bank of clouds building. Not mentioning this to Ruth, I took out my little folding shovel, the same kind I had used in the army, and dug a trench around the tent.

Ruth crawled into the tent first and snuggled up in her sleeping bag. I stayed up a little longer, watching the storm build in the west. Soon I could see the occasional flash of lightning, but the storm was too far away to hear thunder.

I hadn't been asleep for more than an hour when a loud clap of thunder awakened us both. A flash of lightning lit up the tent like daylight, and then the rain began falling, and once more I heard that wonderful sound of raindrops on canvas. The thunder and lightning moved on to the east, but the rain continued to pour down for much of the night. We slept through most of it.

At dawn, although the trees were dripping water and there was mud everywhere, the inside of our tent was completely dry. Our camping neighbors only a couple dozen yards away from our campsite had fared less well. When I asked their little boy about the rain, he told me, "A river ran right through our tent, in one side and out the other." And indeed it had. The young couple was busy gathering up wet clothing, wet sleeping bags, and wet air mattresses and dragging them out of their tent to dry in the bright sun. Later we heard that it had rained more than two inches that night. I was concerned that Ruth's desire for camping would end after just one night. When we were packing up our gear later that afternoon, she said, "Kind of fun to spend a rainy evening in a tent." In the years to come Ruth and I, and eventually our three children, would spend many nights in that used old umbrella tent held up by a big wooden pole.

# One Apps-Power

In the summers when I was a kid, we fished from shore with long cane poles, heavy green fishing line, red and white bobbers, and earthworms. Depending on the month—fishing was usually good in June, less so in July, and not good at all in August—we caught considerable numbers of fish, usually bluegills, sometimes perch, maybe a sunfish or two. But I knew we could do better if we fished from a boat.

We had no money to buy a boat. Besides that, Pa felt that only city people owned boats. Secretly, I knew he would have liked a boat. I knew he enjoyed fishing from boats, as we occasionally did in a rented boat on Norwegian Lake. But Pa would hear none of it when one of my brothers or I suggested he buy one.

Many years later, in 1962, Ruth and I moved with our baby daughter, Susan, from Green Bay to Madison, where I had a new job, with a small increase in income. I quietly suggested to Ruth one day that I would like to buy a fishing boat, not a fancy powerboat like the ones we saw roaring around Lake Mendota, but a fishing boat. I wouldn't even need a motor—just a sturdy pair of oars. She didn't object, so I was off to the nearby Sears store. I found just what I was looking for: a twelve-foot aluminum boat, small enough so I could carry it on the top of my car and thus needing no trailer.

I couldn't wait to show Pa my purchase. The next weekend we drove out to the home farm west of Wild Rose. I showed Pa the boat and the oars, and he asked, "Where is your anchor?"

I said I didn't have one. "Well, you can't fish without an anchor," he said. He dug out an old worn-out plow point to which he tied a length of rope that had been hanging in the barn. We dug some earthworms in back of the chicken house, just as we had done years earlier, and we were off to Norwegian Lake for an afternoon of fishing. Arriving once again at the Anderson farm, we gave them a dollar as we always had for boat rental, but this time we said all we needed was the use of their pier.

I rowed out to the old marl hole. The aluminum boat was much easier to row than the old wooden ones, and for the entire afternoon neither Pa nor I needed to bail water as we did when we used the Andersons' leaky boats. The homemade anchor worked well, the fish were biting, and that evening we all sat around the kitchen table, Ma, Pa, Ruth, baby Sue, and me, eating fried bluegills as only Ma could fry them. When we left for Madison the next day, Pa asked when I might come again, for he couldn't recall when the fishing was any better, and my new aluminum boat had taken his fancy.

Back in Madison, when friends saw my boat and asked me what kind of motor I had, expecting me to say it was a five-horsepower or a seven-and-a-half-horsepower or even a ten-horse, I answered with a straight face, "It's a one Apps-power motor." Most knew right away that meant I rowed it myself, which I did until a few years ago when I bought an electric motor. I keep the boat in my shed at the farm, and every time I walk by it I remember the good times fishing with my dad and later with my three kids as they learned how to fish, learned how to row, and learned how to boast that the boat was powered by a one Apps-power motor.

# They Always Tip Over

One time when Pa and I were fishing in my aluminum boat, I suggested that it would be fun to own a canoe.

"Oh, you don't want one of those," he said. "Canoes ain't safe. They always tip over."

I was sure Pa was wrong, but I knew better than to argue with him when he had made up his mind and had a definite opinion. Still, I believed I had evidence on my side. I had read about the use of canoes by Native Americans, French traders, and early Wisconsin explorers. I had found no mention that any of them considered canoes unsafe.

A couple years later, in the summer of 1965, Walt Bjoraker, my department chair and boss at the university, invited me to accompany him and his two boys on a fishing trip to northern Wisconsin. He said we'd be sleeping in a tent and would have both a boat and canoe for our use. He asked if I would mind fishing with one of his sons in his canoe, a seventeen-foot aluminum Grumman, while he fished in the small rowboat with his other son. I told him I had never fished from a canoe, indeed I had never even been in one. I mentioned that my dad said canoes weren't safe.

Walt laughed. "Unless you do something crazy, like stand up or lean too far over the side, a canoe is perfectly safe."

The morning after our arrival at the campground on the shores of a beautiful little lake in the Hayward area, we got up early, ate a quickly prepared breakfast, and were off fishing. Walt's older son, Gary, was to be with me in the canoe; Walt and his younger son, Gordy, were to fish from the rowboat. Gary, who was probably thirteen or fourteen, had considerable experience with their canoe. So Gary took his place in back of the canoe and I got in the front. At the time I didn't know the paddler in the stern was responsible for steering the craft as well as providing half of the power. Walt demonstrated how to hold and use the wooden paddle properly. He reminded me that I shouldn't use it like an oar. A boat oar is horizontal to the water and is attached to the boat by an oarlock; a paddle is not fastened to anything but the paddler who is using it. The canoe paddle is generally held vertically in the water and can be used on either side.

This seemed simple enough. Soon, with Gary doing the complicated work of steering the canoe and me in the front paddling as I had been instructed, we were off across the lake to a spot where Walt had indicated we might hook a northern pike. I know northern pike, and I knew we weren't talking about a small fish. Many are eighteen to twenty-four inches long and weigh three or four pounds. Some are thirty-six inches long or longer and weigh twenty pounds or more. I wasn't too sure I wanted to be in a canoe with a fish that was longer than the canoe was wide. But I didn't say anything, for in all my years of fishing I had never caught a large northern pike in summer. In summer we mostly caught bluegills, sunfish, the occasional bullhead, and, if we were lucky, a largemouth bass that could weigh five pounds and put up a considerable battle.

We paddled to a quiet little cove where lily pads grew everywhere and where there was not so much as a ripple on the surface. We tucked our paddles away in the canoe and took up our fishing rods. I was using a double-jointed lure that floated on the surface but dove when I reeled in the line. The faster I reeled, the deeper the lure ran. On my first cast I tried reeling in the lure slowly, which was supposed to make it resemble a wounded minnow skipping along on the surface. The second time I cranked the reel a bit faster, and the lure ran just beneath the surface. I saw nothing following the lure either time, not a small fish, not a large fish—no fish at all.

The third time, I cranked the reel slowly at first and then stopped. I cranked it more rapidly, sending the wiggly little lure deep into the water; then I stopped and allowed the reel to return to the surface. Not expecting anything, I was slowly reeling when it happened. The water around the lure exploded, and within the geyser of water, I saw the fish, a northern pike that I made to be at least twenty inches long. I set the hook and began fighting the big fish, all the while conscious that I was in a canoe. Pa's words quickly came back: Canoes ain't safe. They always tip over.

Gary, now with his paddle in hand, was working to hold

the canoe steady as I continued fighting the big fish. It leaped out of the water, shook its head trying to shake loose the lure, then dove deep and tightened my fishing line as tight as a well-tuned guitar string. My fishing rod bent half over, and I wondered if the line would hold as I continued to alternately crank on the reel and allow the line to play out depending on what the big fish was doing. I wished I had paid more attention to the drag feature on my reel, which allowed a fish to strip line from the lure, but with difficulty. The drag feature was designed to eventually tire a fish so it could be brought to the boat.

Finally I had the big northern alongside the canoe. I wasn't too sure I wanted to lift it into the canoe, for northern pike have sharp teeth, and a freshly caught northern flopping around in the bottom of the canoe wasn't something I wanted to contend with. Gary suggested that we paddle to the nearby shore, remove the fish from the lure, and put it on our fish stringer—a length of light cord that allowed the fish to trail in the water outside the canoe.

With the northern on the stringer, and a good story to tell when we arrived back at camp, we continued fishing, catching a few good-sized bluegills but no more northerns.

When I returned home from the fishing trip on Sunday afternoon, I was convinced that I wanted a canoe. I now knew they were safe if you behaved yourself. And you could use them for fishing. The following year, again with Ruth's approval, I bought a seventeen-foot Grumman canoe, which I own to this day. It's a bit heavy, but it fits nicely on top of my car. It is scratched and scarred from years of bumping into rocks and scraping over gravelly landing places. But it has never failed.

We have two small ponds at our farm, Roshara, and for several years we stored the canoe on the bank of what we call Pond I. It was there for the kids to use whenever they wanted to tour the nooks and crannies of this little body of water. With my father's admonition in my head, before I would allow the kids to go out in the canoe I instructed them to put on life jackets and take the canoe out into the middle of the pond and try to tip it over. It was far more difficult to do than they—and my dad—thought. But by going through this exercise they knew what they could safely do with a canoe.

I can think of only one canoe accident. My son Steve, the photographer in the family, was alone in the canoe, taking pictures of water plants. He had no counterweight in the front, which allowed the front of the canoe to tip up out of the water. As Steve leaned over the side to snap a photo, the canoe tipped, and Steve learned a valuable lesson about centers of gravity.

Another time, Sue and Steve were canoeing on a bright, sunny day when the water on the pond was smooth and paddling was easy. The canoe had been pulled up on shore and had not been used for several weeks, and a family of garter snakes had made a home in one end of the canoe behind some of the flotation material. With the canoe well out in the pond, the snake family, unaccustomed to the sound of paddling and the movement of the canoe, began emerging from their newly found home. The strategy the kids agreed on, once they got over the shock of several snakes slithering around in the bottom of the canoe, was to slip a canoe paddle under each snake and toss it in the water. This they did until every last snake was out of the canoe.

In addition to all of the fun and adventures the kids had with the canoe on the pond, I have found the canoe to be one of the best possible ways to quietly explore every little corner of the pond. I've also enjoyed just letting the canoe quietly float along where it will. My canoe is one more way to connect to water, to be close to it and yet not in it. Canoes are minimally intrusive. And they are rich with the stories of earlier days when Native Americans, explorers, missionaries, and others used canoes as their prime means of transportation.

# Mecan River Canoe Trip

In 1977, when Steve and Jeff were thirteen and fourteen, respectively, they wanted to go on separate camping trips with me. I spent a couple of nights camping with Steve at Yellowstone Lake near Blanchardville, where we did some fishing and boating.

Jeff wanted a canoe trip on a river. There were many rivers near us to choose from, but I eventually picked the Mecan, a relatively narrow river that wanders through southwest Waushara County and northwest Marquette County in central Wisconsin. Traveling to our central Wisconsin farm on Highway 22, we pass over the Mecan a few miles north of Montello.

Looking at the map, I located the Germania Marsh Wildlife Area, which the Mecan passes through on its way to the Fox River. The Germania Marsh comprises some 2,400 acres in north central Marquette County—and I guessed there must be an island or some high ground where we could set up our tent and spend the night before continuing on our journey.

On a Saturday morning in mid-October, one of those cool, bright sunny days when the trees are ablaze with color, Jeff and I loaded our seventeen-foot Grumman canoe on the top of the car. We packed lunch, supper, and breakfast in our cooler, gathered up my tent (by this time I had a small two-person tent that was easy to put up) and our sleeping bags and air mattresses, and we were off.

At that time, in the late 1970s, the Mecan was not yet a popular canoeing river. I didn't know this when we put in our canoe, loaded our camping gear and food, and floated off. The current was quite brisk—we had only to keep the canoe in midstream and the current moved us along at a brisk pace. After the first mile or so I relaxed. I sat in the back, doing the steering, and Jeff paddled in front. Neither of us had to work very hard, so we spent most of our time looking at the fall colors and commenting on what a fine day it was and what a great canoe river we had found—especially because we apparently had it all to ourselves. I noticed several sandhill cranes flying overhead, and I pointed them out to Jeff. I knew that at this time of the year the sandhills were congregating in preparation for their annual migration south.

For the first hour of canoeing, we saw not one other canoe. As it turned out, we would not see another canoe for the entire two-day trip. Rounding a bend in the river, we saw why.

"What's that, Dad?" Jeff asked. He had stopped paddling and looked back at me.

"Just a tree," I said quietly. But I could see that the tree, probably toppled by a windstorm, stretched all the way across the river, making passage impossible. I had brought along a hatchet for cutting campfire wood, and now I chopped away at the downed tree, cutting off limbs and tossing them aside, all the while staying in the canoe and trying to keep it from tipping as the current insisted on pushing it farther into the tangle of the downed tree.

Once I had removed many of the limbs, I suggested that Jeff climb up onto the downed tree trunk, and then I did as well. We dragged the canoe over the tree and gently let it down on the other side, climbed back into the canoe, and were on our way. It had taken us nearly a half-hour to move beyond the blockage. Jeff saw all of this as part of the adventure. I just hoped there were no more trees to block our way.

But there were more trees, three or four more that we encountered, again chopping off branches and pulling the canoe over them, before we finally arrived in the Germania Marsh, a wetlands area where the river twisted and turned and slowly made its way through tall grass. I asked Jeff to keep an eye out for an island or any dry spot where some trees grew and where we could pitch our tent for the night. Finally we spotted a little island, maybe a half-acre in size. We pulled up our canoe as the October sun was slowly sinking in the west. My arms ached. My back ached. My aches were not from paddling but from chopping tree limbs and dragging our canoe over downed trees.

Jeff helped me set up the tent, and while he was digging in the cooler for the hot dogs and buns, I scouted for dry firewood. Soon I had a campfire going and we were roasting wieners as we watched the sun set in a brilliant red sky. The yellow flames of our campfire reflected off the sides of our tent, and a trickle of wood smoke drifted off to the east. I lit my pipe, sat with my back against the canoe, and relaxed.

"Kinda fun, isn't it?" Jeff said.

I agreed with him, and I resisted the urge to say something about my sore back and aching arms.

We stayed up watching the campfire and the stars overhead, smelling the primitive smell of campfire smoke, and I told Jeff stories of the times I had spent huddled over a smoky campfire when I was his age and ice fishing. Then it was off to bed, into our warm sleeping bags spread out on air mattresses.

Not long after we were in bed, the sandhill chorus began. It surrounded us, the prehistoric call of these big birds echoing on the night air. Unbeknownst to me, Germania Marsh was a major gathering place for sandhill cranes before their fall migration.

"Dad, I can't sleep," Jeff said. "The cranes are too noisy. Why are they making so much noise?"

"Oh," I said, "crane families that haven't seen each other all summer are back together. They're talking about their youngsters that hatched in the spring and how well they are doing. They are talking about the place where they nested and how that worked out for them. They're talking about their upcoming trip south and what it will be like this year."

"Right," Jeff said, pulling his sleeping bag over his head.

Finally we did get to sleep. We continued our canoe trip the following day, this time without the inconvenience of chopping limbs and dragging our canoe over downed trees.

But even with the unexpected happenings—downed trees and loud sandhill cranes keeping us up at night—this two-day canoe camping trip made our list of most memorable adventures. A river is a wonderful place to get to know a son or a daughter better.

# Whole Pail Shower

When we bought our central Wisconsin farm, the original house had long ago burned down and the remaining farm buildings were in bad repair. The old never-painted barn was leaning to the south, the pump house was scorched in the fire that took the house, and the granary/wagon shed/chicken house combination was within a few years of falling down.

The place had no electricity, and its water source was a hand-operated pump in the dilapidated pump house. The first couple of years after we bought it, when the kids were still little, we went there during summer weekends and vacations, living in a tent while I worked on making the granary livable. One of the first things I did was to have the electric co-op string a line to the granary and to the pump house. I bought the pump jack and electric motor that had pumped water at the country school just down the road and, with the help of a neighbor, got the pump working so we had water—excellent water, as it turned out.

One day our neighbor across the road, Floyd Jeffers,

came over to see how we were coming with our attempts to turn a 1912 granary into a cabin and to share a story or two about this community that at one time had been known as Skunk's Hollow.

I showed him the working pump and the little washing station I had set up at the west end of the pump house where we could wash off the dirt and grime after working around the cabin.

"You realize, Jerry," Floyd said upon looking at the pump, "this is some of the best water in all of Wisconsin."

"Glad to hear it," I said.

Later, when I was researching the geology of our area, I discovered that our farm, located on the terminal moraine where the glacier stopped, is also within a mile or two of a water divide. The rivers and streams that begin west of our farm flow toward the Wisconsin River and then south to the Mississippi and on to the Gulf of Mexico. The rivers and streams east of our farm flow to Lake Winnebago and then north to Green Bay and on to the Atlantic Ocean. Floyd Jeffers was right about our water's purity and good taste. We are at the very beginning of the aquifer that flows east, where no contaminants have yet found their way into the underground water source.

Starting with the late 1960s, the family experienced what I had known from my childhood days, when water was scarce—we used as little as possible and made every drop count. The pump worked okay, but I didn't want to overwork it, so we made certain to waste not a drop of water that flowed from its pipe.

Rather than have everyone bathe in a big galvanized washtub, as I had done growing up, I invented an outdoor shower for the farm. First, I cut three eight-foot-long by three-inch-thick black locust trees. I formed the trees into a three-legged tripod, which I wired together at the top. Then, back in Madison I went to a local hardware store owned and operated by my neighbor, Maury Ellis. I told Maury I needed a fourteen-quart pail, a length of garden hose, and a sprinkler head.

"That's a rare combination," Maury said.

"Here's what I want," I continued. "Can you cut a hole in the bottom of the pail, fasten about two feet of garden hose to where you made the hole, and then attach the sprinkler head to the other end of the hose?"

"I can do that," Maury said. "But what am I making?"

"An outdoor shower for our farm," I answered, going on to explain how I would tie a rope to the pail's handle, and then thread the rope through a little pulley I had fixed to the top of the black locust tripod. I could thus lower the pail to fill it with water, and then pull it to the top of the tripod, so that the sprinkler head would be just above a person's head. The sprinkler head had a little handle that would turn the water flow off and on.

I installed the water pail with its hose and sprinkler head, and the outdoor shower was ready for its first occupant. Ruth and Sue said the shower still missed something; they didn't want to shower with the whole world watching. So I wrapped an old canvas around the three-legged shower stall, and now it was ready.

We combined water warmed on our woodstove with cold water pumped from the well to get the temperature just right before pulling the filled pail into place. We also discovered that with careful attention, one pail of water was enough for three showers. Using more than that meant heating more water on the woodstove and carrying more cold water from the pump—strong incentives for using as little water as necessary when showering.

One of the rewards for hard work at the farm—helping me hoe in the garden, helping with brush removal, assisting with cutting the long grass around the cabin, and a hundred other tasks we faced at Roshara—was a whole pail shower. That is, the winner of the prize for hard work got to use the entire fourteen quarts of water for his or her shower.

"Is this a whole pail shower job?" became a popular question at the farm as I assigned various duties to the kids. Many years later, the topic of a "whole pail shower" often comes up when my children try to explain to their children that water was scarce at the farm and it had to be conserved—a valuable lesson passed from one generation to the next.

# Water, Learning, and Creativity

I have taught creative writing for forty-five years, in a variety of settings ranging from the back rooms of tiny libraries to fancy conference rooms at universities. No matter the type of venue, my favorite place for teaching is anyplace near water. Something about proximity to water not only sparks creativity but also has a soothing effect on participants.

For three summers in the 1980s, I taught a creative writing workshop on Washington Island off the tip of Wisconsin's Door County peninsula. The island has a permanent population of 660 residents; this number increases considerably during the summer tourist season. There are no bridges to Washington Island; most people reach it by riding a ferry seven miles across a stretch of Lake Michigan known as Death's Door for the number of shipwrecks in these waters years ago.

The ferry crossing can be an adventure, especially on a stormy, windy day when the waves kick up and toss the sturdy boat around. On one ride to the island, Ruth and I left the tip of the Door Peninsula in a dense fog and a substantial wind—two ingredients that don't often occur together. We weren't more than a mile from shore when the waves began breaking over the front of the ferry, splashing on the cars parked there in neat rows. The ferry bucked like a wild mustang as it plowed through the rough water and dense fog. Ruth stayed in the car, while I decided to crawl the stairs to where the captain and the first mate were operating the ferry. The captain had his face glued to the green glowing radar in front of him; the first mate hovered over the ferry's controls. I overheard the captain talking on the radio with an ore boat crossing Death's Door.

"Do you see any fishing boats, George?" the captain asked.

"Nope, so far so good," I heard the ore boat captain say.

Neither captain seemed concerned about the welfare of their own boats; they feared running over a much smaller fishing boat caught in the fog and the waves. These captains, with many years of experience between them, knew both the wonders of working on water and the potentially devastating dangers.

My island workshop classes were designed for adults who worked in human services—social workers, psychiatrists, psychologists, pastors, priests, and other religious leaders. With weather permitting, I held the classes on the shore of a little bay where the participants could listen to the sound of the waves splashing on shore while watching the sunlight bouncing off the water, creating an ever-changing scene of contrast and color. The workshops were held from Sunday to Sunday, and to my astonishment it would take some of the participants a couple of days before they calmed down enough to do any creative writing.

I especially remember one student, an experienced social worker. During the first couple of writing sessions I could tell that she had trouble concentrating. It was the second day, on one of my early morning hikes, that I found her sitting all alone and looking at the water.

"How's it going?" I asked as a way of making conversation.

"Much better," she replied. "Much, much better."

"Glad to hear it," I said, wondering what was working for her.

"It's the water," she went on. "So calming."

I became a firm believer that working near water helps to speed the calming process as well as encourage creativity. This realization was confirmed when I taught workshops for the University of Alaska in the mid-1990s. For two summers, I taught on Yukon Island, seven miles from Homer, Alaska, in Kachemak Bay at the tip of the Kenai Peninsula. The island is about a mile wide and a mile and a third long, with no electrical service, roads, or automobiles. To arrive at Yukon Island required a forty-five-minute boat trip along a route where mountains come down to the water's edge and few people live.

My students were receiving graduate credit for attending. They flew into Homer from all around Alaska, the North Shore, Fairbanks, Anchorage, and many smaller villages. They brought camping equipment and pitched their tents on Yukon Island's shore.

I conducted the workshops in early July, when the sun rose by three a.m. and it was still light enough to work outside at eleven at night. I have rarely seen such natural beauty as at Yukon Island, with the mountains reflecting in the clear water of Kachemak Bay, and during breaks I often spotted participants sitting at the water's edge, not doing anything but looking at the water and the mountains. I came to understand that the setting had as much to do with their learning as my efforts in the classroom did.

I've observed the same thing at The Clearing, a residential learning center on the shores of Green Bay in northern Door County, Wisconsin. I have taught weeklong and one-day classes there for more than twenty-five years. Students come to learn about writing but also to relax and regenerate, to rediscover a creative self that has been buried in the day-to-day activities associated with making a living. During breaks I encourage my students to walk by the water or just sit and look out across the bay, toward the shore of Upper Michigan, which can be seen in the distance on a clear day.

Students at The Clearing sleep in cabins dotting the property and eat their meals together in a large dining hall that overlooks the bay. One cabin, called the Cliffhouse, is especially remote, situated on a cliff overlooking the bay. There is no electricity in the Cliffhouse, so those who spend the night become acquainted with a kerosene lamp and come away with a new understanding of how dark the night can be. They also gain a new insight into the sounds and sights of water. If the night is windy or stormy, waves pound against the rocky shore, sending up spires of foam and making a sound as old as the earth itself. The sights and sounds of water pounding on rocks, especially for those who have never experienced them, can be a transforming experience.

A Clearing tradition is to watch the sunset each evening over the bay. Students gather on the cliffs near the school-house, where most of the workshops are held. There they sit and watch and say little. Every sunset is a little different from the previous evening's, sometimes dramatically different if a thunderstorm is brewing in the west. It's nearly impossible to describe a sunset with words, and even more difficult to express a sunset's meaning. But a sunset over water enthralls those who experience it, and people often come away with a new appreciation for one of nature's profound moments, one that's too often taken for granted.

Watching a sunset over water or listening to the waves pounding against a rocky shore can awaken the creative juices in each of us. In turn these experiences calm us down so that newly awakened creativity can be expressed—in words, in paintings, in wood carvings, in music, and in whatever artistic expression interests us. Water can speak to us. It can touch our innermost being. Water can contribute toward making us more human.

# Piscatorial Retreat

I called it a piscatorial retreat. I thought it sounded academic, like something important for graduate students to do, perhaps even a new type of learning experience. The retreat, which I offered nearly every year in the 1970s and 1980s to my University of Wisconsin graduate students majoring in adult education, was a celebration of the opening day of fishing season.

My students and I spent a weekend fishing on the Peshtigo River in northern Marinette County, not far from the Upper Michigan border. We camped at McClintock Park, which offered campsites and outhouses and a spot for a campfire on the banks of the Peshtigo. The fishing opener in Wisconsin is usually the first Saturday of May, and the river teemed with trout and roared with the snowmelt. We wore rubber waders that came up to our chests, which allowed us to be in the river—or as one of my graduate students said, "To be one with the river."

The water was cold, forty degrees or a little warmer. If we were lucky, those early days in May would be warm and sunny, with temperatures climbing into the seventies. Another student, upon emerging from the frigid water after an hour or so of fishing, said that he had learned the true meaning of *average*. The lower half of his body was freezing, the upper half was warm, so on average he must be comfortable.

We caught trout—cold-water trout, rainbow trout, tasty trout—and prepared them in the evening on a camp stove. My students were from many parts of this country and even Canada. A couple of them had never fished before, and the majority had never spent a night in a tent. Several of us slept in the old umbrella tent I had purchased when I lived in Green Bay.

Every evening after dinner, we built a campfire, and as we watched the yellow flames licking at the pinewood, we listened to the river. There was a waterfall just a few hundred yards upstream, alive with water pounding on the rocks and sending threads of water high into the air.

As we sat watching and listening to the rushing water, we swapped stories of fishing in Canada and in North Dakota, and I got to know my students in a way I never would in the classroom or my university office. I suspect they got to know each other, and themselves, in a different way as well.

One year the weather did not cooperate. As we traveled the many miles from Madison to Marinette County, the afternoon was cool and cloudy. We arrived at our campsite just before dark. With the tents in place we started a campfire, which warmed the night, provided light, and sent us to bed with enthusiasm for the morning and opening day.

I had convinced a colleague, Professor Pat Boyle, to accompany us on this trip. He had never fished for trout or slept in a tent. We all had warm sleeping bags, so we slept comfortably enough. I woke up once during the night to find that the inside of the tent was frigid and the tent roof appeared to be sagging. My first thought was that I hadn't tightened the tent's guy ropes well. I had, in fact, but looking outside I saw that it was snowing—hard. The snow was causing the tent roof to sag. The next morning I saw that it had snowed three inches overnight.

When we crawled out of our tents at sunrise, we were greeted with a return to winter and a challenging day for fishing the Peshtigo. I immediately started the campfire, and soon everyone was clustered around it, warming their hands, front sides, and backsides. We scraped the snow from the picnic tables and prepared a lukewarm breakfast using my sputtering propane camp stove, which didn't work well in cold weather. My students had come to fish, and fish they did, some from shore and some braving the current and cold water of the Peshtigo. I pulled on my waders and joined the group in the river, not especially comfortable, but not freezing either. I did have one small problem, especially in the morning when the temperature stayed below freezing: the eyelets on my fishing rod froze, and I continually had to break the ice from them so my fishing line would move through them. Fishing was slow. I caught one or two trout, as did several others.

Pat Boyle later told me that fishing when it was that cold was the dumbest thing he had ever done. He spent the entire day sitting by the campfire, except for the time he spent gathering more wood. I told him that afternoon that he smelled like a smoked salmon.

"At least I was warm," he said.

Although that particular fishing trip had its challenges, it was one that my graduate students never forgot. A graduate student from North Dakota later told me, "Jerry, thank you for asking me along for that snowy weekend on the Peshtigo River." He continued, "There's reading a book, and there's listening to classroom lectures, but sitting around a campfire with the music of the river in the background is something I never forgot. It reminded me that I need to slow down more often and appreciate the little things—like the sound of a river in the night. And that there is more to learning than academics."

There is just something about a river, especially a fast-moving, noisy river like the Peshtigo, that provides unique learning opportunities.

# Boundary Water Adventures

Tucked up against the Canadian border in northern Minnesota is one of the world's special places, especially for those of us who enjoy rivers and lakes and pure, sparkling water. The place is commonly referred to as the Boundary Waters; its official name is Boundary Waters Canoe Area Wilderness (BWCAW). According to the USDA Forest Service, the BWCAW is over a million acres and extends nearly 150 miles along the international border. This is the land of the French voyageurs, who canoed these waters more than two hundred years ago, trapping and trading with the Native Americans. The area was set aside for recreation in 1926 and became a part of the National Wilderness Preservation System in 1964.

It is a land of loons and moose and wolves that howl in the night, if you are lucky enough to hear them. But mostly it is a land of lakes and rivers, with more than a thousand lakes that vary in size from ten to ten thousand acres, plus hundreds of miles of rivers and streams. Water makes up about 20 percent of the BWCAW surface area—which means one

can travel quite easily by water. But the travel must be by canoe or kayak, as no motors are allowed in most of the region.
My son Steve and I have visited this place since 1983. At times my son Jeff and daughter, Sue, have also joined me there. When people ask me why I do it—why make the nearly eight-hour trip from Madison to northern Minnesota to paddle a canoe, sleep in a tent, give up all modern conveniences, and warm myself by a smoky campfire with a sliver of moon coming up over the dark waters of a lake—I simply reply, "Because the experience is special." What I really want to say is, "Because it's good for my soul to have it enriched once a year." But I feel many people wouldn't know what I was talking about. And that would be understandable, because until you've been to the Boundary Waters, you cannot have that deep, profound feeling that only the direct experience can provide.

For every Boundary Waters canoe trip, I've kept a detailed journal. I do this to remind me of all that I did and saw, but also because when I write about an experience, I often find that a deeper meaning emerges in the process.

*Friday, August 17, 1984, 7:30 p.m., Loon Lake*

Supper time tonight was dried hash brown potatoes, summer sausage, and dried applesauce. I paid $1.95 for the applesauce and it was terrible. The worst. After supper Jeff and I took the canoe out on the lake for a couple hours of fishing, but we hurried back when we heard thunder. Now, back in camp, Jeff is writing in his journal and Steve is building a campfire. The thunder is coming closer.

What a beautiful place, from the loon calls echoing across the lake to the magnificent firs and white pines and the blue-green water of the lake. The boys are excellent campers, willing to share in toting wood, carrying water from the lake, and not grumbling at dishwashing time. The two of them take turns portaging the canoe. By camping in one place and then making day trips from our permanent campsite, the portage loads, aside from the heavy 17-foot Grumman canoe, are light.

The campfire is snapping and crackling as the thunder booms in the background. Very quickly it has become too dark to write. But I must take time to jot down a few lines about my feelings about being here in the Boundary Waters, at this isolated campsite, on a beautiful lake. First of all, it gives me a chance to be alone with my two sons without any interruption—everyone's life is full of interruptions these days. Secondly, it is the sheer beauty of the place. Tonight, thunder rumbles in concert with the loon calls, a most interesting set of contrasting sounds. Both so old and primitive that it's difficult to imagine their beginnings.

### Saturday, August 18, 1984, Loon Lake

We are up at 6:30. Last night the storm clouds that darkened the sky and thunder that grumbled and rumbled presented us with only light rain as we crawled into our tent. But sometime during the night, the rain turned into a downpour, waking us up to check the tent windows. We soon fell asleep again, to the steady pounding of the rain on the tent canvas—an altogether pleasant sound.

*July 16, 1991, 9:10 a.m., Pine Lake*

Steve and I just finished a breakfast of pancakes, the shake and bake type, which really hit the spot. A light breeze from the southwest is keeping the mosquitoes away from our campsite, which is on a rocky outcropping facing west.

Yesterday, a fierce wind blew all day, spreading whitecaps across the lake. We remained in camp, patiently waiting for the wind to subside, but it never did. Around midafternoon we tried to launch the canoe for some fishing, but we failed. We couldn't push off without waves washing over the sides of the canoe. At one point in our effort, after a huge wave wet Steve's backside, he yelled, "Abandon ship" as he scampered out of the canoe onto the nearby rocks. In the early evening the wind began subsiding, and the waves turned into a light chop. We launched the canoe and paddled a mile or so up the north shore of the lake, but no fish.

The wind continued howling all night, splashing waves against the rocks below our campsite. It was a great night for sleeping.

I enjoy lakes and rivers wherever they might be, but there is something special about being in a place where no cottages or fancy million-dollar lake homes crowd against each other all the way around the lake, a place where noisy, high-powered motorboats don't churn this way and that, interrupted by the even noisier jet skis that skim the surface, making tight turns and sending up huge plumes of water.

The Boundary Waters is a quiet place. A place for contemplation. A place to leave behind the noise and bluster and pressure of our hurry-up society to discover a bit more of who we are as people and consider our place in the cosmos. A place to be one with nature.

# Missouri River Challenges

On a warm, muggy day in mid-June 1992, we put fifteen rented canoes, with two people in each, into the Missouri River at the Garrison Dam, some sixty miles north of Bismarck, North Dakota. Spring rains had filled the river to capacity, and although it wasn't flooding, the current was strong and a bit challenging for canoeists with little experience.

I was leading a group of higher-education middle managers from around the United States who were enrolled in a leadership development program I directed. The previous week we had lived with the Mandan Indian tribe on their reservation. Now we would canoe south, camping along the Missouri River on our way to Bismarck.

All of the participants had advanced graduate degrees. They were departmental heads, assistant deans, or the like, and all aspired to a higher-level position: university president, college dean, or director of a higher education program. Except for two or three of them, they had little or no experience canoeing and even less experience tent camping.

Those of us with canoeing experience spent an hour or so before launching giving basic instructions—with the strong current, understanding steering was even more important than paddling, as the canoes would travel downriver with little help. I insisted that each person wear a life vest and that the canoes stay together so that no one got lost among the Missouri River's many islands, branches, and dead ends. Traveling with us were a river guide with years of experience guiding canoeists on the Missouri and a Mandan shaman whom the group had met during their stay on the reservation. We were to camp that night at a location that had been a campsite for the Lewis and Clark expedition that had traveled this way in 1803.

The first day went well. With the strong current we reached our first destination by midafternoon, set up our tents, and relaxed on the banks of the rapidly flowing Missouri. By day's end almost everyone in the group had learned enough about paddling to be fairly comfortable on the water. But muscles were sore, and I heard a few people complaining about what paddling a canoe on a river had to do with leadership development. In case of emergencies, I had one of my staff people following our progress in a car. He ate meals with us and kept track of weather reports. We talked regularly via handheld radios.

Upon arriving at our first campsite on that first day, one of the participants, an assistant dean, had absolutely refused to sleep in a tent. Furthermore, she said that what we were doing had nothing whatever to do with her developing the administrative skills she needed to become a top-level higher education officer. She insisted on sleeping in a motel. I didn't argue, but I did say that as far as I knew there were neither towns nor motels within forty miles of our campsite. A couple of her fellow participants put up her tent. She watched them, grumbling the whole time. One or two others questioned what we were doing and how it related to leadership development, but they were willing to give the experience a try.

A full moon greeted us that evening as we sat by a blazing campfire and listened to the sounds of an angry Missouri River hurrying south, at times tearing at the riverbank. I had asked the shaman, who had a master's degree in sociology, to share his people's creation story. He had a great sense of theater. Dressed in fringed buckskins, he stood by the campfire with the moon rising over his shoulder and the river in front of him. He explained how important water was to his people, saying that for generations they had traveled on it, fished in it, and depended on it for their livelihood.

Most of the participants were accustomed to the "talking points" approach to giving presentations, moving from one topic to another in linear fashion. The shaman was a storyteller who followed a nonlinear approach. I could tell by watching the faces of the participants that several were bothered by his approach. When he finished his story most of them clapped, but not all. A few students told me later that busy administrators simply didn't have enough time for telling a story. Today's and tomorrow's leaders must move beyond such old-time approaches, they said. I didn't argue with these naysayers, for I knew that unless they discovered for themselves the importance of storytelling as a form of communication, they wouldn't use it.

The majority of the group, however, was enthralled by the shaman's storytelling, especially because the stories were about the river and its importance to his tribe.

With the campfire coals still glowing and the full moon bright overhead, the tired, muscle-sore campers crawled into their tents and were soon sound asleep, the sound of the river gurgling in the background.

After breakfast the next morning, we were back on the river on our long journey south. As I paddled, I noticed a bank of clouds building in the west. After an hour's break for a shore lunch, we were back paddling, enjoying the river. The bank of clouds continued to build. Just as I was about to radio my man in the car, he contacted me. "Near hundred-percent chance of strong thunderstorms this evening," he said.

I told him I had seen the clouds building and wondered what kind of weather was headed our way. We found our scheduled stopping place, another Lewis and Clark campsite, by late afternoon. Several of us pulled the canoes out of the water, turned them over, and tied each one to a tree. I told the campers that a storm was headed our way and that they should pay particular attention to how they pitched their tents. "Make sure you pound in all of the stakes, and fasten the guy ropes leading out from the tents to something secure—a small tree works well," I told them. I also suggested the group not pitch their tents under a tree with dead branches, for fear a strong wind might send a branch crashing into a fragile tent, and I warned them that if a storm came in the night, they should stay in their tents, where they'd be safer than in the elements.

I'd noticed the previous evening that several of the campers had put in only a few stakes, and some had avoided using the guy ropes entirely. With clear weather that didn't matter. In a rainstorm it would. Months before the trip, I had carefully described what kind of tent to purchase or rent for the canoe trip. I told people to avoid the big, clumsy, rather tall family tents. "They won't stand up to heavy weather," I'd said. But several people hadn't listened to my suggestions, and there were three big, clumsy family tents, the kind in which a six-foot-tall man can stand up without bumping his head on the canvas top, set up among the lower, more sturdy and weather-worthy tents.

That evening we once again sat around the campfire with the shaman telling stories. Clouds obscured the moon, so a few feet away from the tent it was as dark as the inside of a cave. The air was still. There were no bird sounds, no sounds at all except the subtle sizzling of the dying campfire and the Missouri River racing south.

By ten p.m. everyone had gone to bed. Those canoeists who were not accustomed to so much physical exertion were tired and sore. I stayed up a bit longer, sitting by the glowing embers of the campfire and listening to the river—how I enjoyed that sound. I thought about Lewis and Clark making this journey nearly two hundred years earlier. I wondered whether they sat where I was sitting, listening and thinking of early days and what was to come. I wondered whether the river held meaning for them beyond merely serving as a transportation route.

I heard it first as a low growl off to the west. Looking in that direction I saw jagged flashes of lightning cutting across an angry sky. I guessed the storm would be upon us in an hour or so. I went off to my tent, but I kept on my clothes and shoes and stayed awake. I knew that North Dakota spring storms can be fierce, with heavy winds and downpours of rain.

The claps of thunder grew louder and the lightning flashes brighter and closer together. Thunder rolled down the river, on top of the water, echoing throughout the valley. And then the rain started, at first a few big drops pounding on the canvas top of my tent, then more, like a drummer starting slowly and moving more quickly, and then I couldn't separate the sound of one drop from another as the clouds dumped their heavy load of water on our campsite. I heard a few yells as people woke to the noise, and I glanced outside for fear that someone might be walking around in the storm. No one was. Apparently they had heeded my advice to stay inside their tents in the event of a storm.

The wind came up, a strong, straight wind that furiously shook my little backpacking tent, but the tent stood strong, as it had in the many previous storms it had withstood when I canoe-camped in the Boundary Waters of northern Minnesota.

A brighter than usual flash of lightning with a simultaneous clap of thunder shook the ground under my tent, and for an instant everything inside and outside was as light as day. I knew the lightning strike had been close to our camp, and I hoped no one had been injured. I heard no yelling or screaming, a good sign. The pouring rain continued.

After a half-hour or so, the rain stopped, the wind subsided, and the storm moved on down the river. The sound of the grumbling thunder slowly disappeared, and the only lightning was a few distant flashes to the south.

Grabbing my flashlight, I unzipped my tent flap and began walking through the camp. People were coming out of their tents, some of them dry and smiling, more of them wet and grumbling. All of the big, family-type tents had collapsed. One tent frame was bent enough to render it useless. Thankfully, those in these big tents were not hurt, only frightened and soaked.

Those who had not properly staked and tied down their tents also found their shelters collapsed around them. Some campers had pitched their tents in natural waterways, and now rainwater rushed through their tents, soaking them, their sleeping bags, and their clothing.

There were a few bruises from tents collapsing, but outside of a good scare and a thorough soaking, everyone seemed okay. I suggested that those with downed tents crawl in with those people whose tents were still standing, and we'd do a more thorough assessment of damage in the morning.

The next morning was clear and bright with sunshine and a warm breeze coming from the south. I checked the empty can I'd left outside my tent and guessed we'd gotten at least two inches of rain. I found some dry wood under the woodpile from the previous evening and with some difficulty managed to start a campfire. Campers slowly found their way toward the campfire, hoping to dry some of their wet clothing.

I made a tour of the camp, checking on the torn and broken tents and wondering if we had enough good tents left to continue our journey down the river.

Just fifty yards or so from the nearest tent, I found where the lightning had struck an old pine tree, tearing out pieces of pinewood and scattering them all about. Had the tent been any closer, the jagged pieces of wood, sharp as daggers, might have torn through it and severely injured the occupants.

As everyone gathered around the campfire, some smiling and some frowning, all looking more than a little sleepy, we discussed whether we should continue on or abort the trip. About two-thirds of the tents were still usable.

About half of the group wanted to continue on, the rest said no. I decided not to go on. I contacted my staffer with the radio and asked him to arrange for a bus to pick us up. After a couple hours' wait for transportation and a bumpy school bus ride, we returned to a hotel in Bismarck. Dirty, wet, and disheveled, we gathered in a conference room and discussed the canoe trip. It was too soon for many of the participants to fully comprehend what they had learned from the couple of canoeing days on the river, especially one that ended with a storm that had raised havoc and soaked nearly everyone.

One interesting comment I heard was, "Being soaked in the rain makes everyone the same." How true, I thought. By dint of position or the prestige of their institution, some in this group seemed to think they outranked others. But when everyone was wet, cold, and miserable, a sameness of circumstance had settled over everyone. Someone said how grateful he was to several fellow campers who were suffering just as he was but had been ready to help him in any way they could.

I said little during the discussion, for the river and the storm had done more teaching than I could have dreamed of doing. The river was teaching these future academic leaders the meaning of compassion, the importance of helping each other, and the acceptance of people as they are as fellow humans, without first thinking about degrees, titles, and institutional affiliations.

As the months passed, I heard from many of the Missouri River canoeists who shared with me that our river trip had been one of the most powerful events in their lives. Many said how much they had learned about themselves. Several shared that they had now a very understanding of what leadership meant. Leadership fundamentally is about caring and sharing and, above all, concern for the people around you not as followers but as fellow human beings. Our journey down the Missouri River and our miserably wet night seeking shelter together had taught them more than any leadership book or classroom lecture ever could.

# Family Gatherings at the Lake

It began in 1971, the first year that I taught creative writing at the School of the Arts in Rhinelander, sponsored by the University of Wisconsin–Madison. In lieu of payment, I asked to bring the family along. We would stay at one of the many resorts in the area. Oneida County, where Rhinelander is the county seat, boosts some 428 named lakes.

Our kids were nine, eight, and seven, and none of us had ever stayed at a lake resort. The cabin assigned to us was just a few feet from the lake. But the cabin had seen better days. In the bedroom where Ruth and I slept, we wondered why the bed was stuck against the wall nearest to the lake. We pushed it back into place—and then slowly watched as the bed rolled back to where it had been. The cabin was leaning rather significantly toward the lake.

"Will we end up in the lake?" Ruth asked. I assured her that we wouldn't. We didn't tell the kids about the tipping cottage; they were having too much fun getting acquainted with the lake, where they could swim every day while I was teaching at the school and go fishing with me in the evenings.

Spending two weeks in a tipping cottage in Oneida County proved the highlight of the kids' summer adventures, and Ruth and I enjoyed it as well. We loved the sound of the waves lapping on the sandy shore, the sunset over the lake in the evening, the wild ducks swimming by. I found it relaxing, and it also sparked my creative juices, as every evening I not only prepared for the next day's workshop but found myself coming up with new writing ideas of my own.

I taught at the School of the Arts for thirty-two years. The kids joined us until they were in high school and had summer jobs. For several years it was just Ruth and me. By 2002 the kids were married with kids of their own, and the whole bunch wanted to come to the resort for a week. Our kids were reliving a bit of their childhoods and sharing with their kids their experiences of many years earlier: water-skiing, swimming, fishing, listening to the loons call, watching little ducks swim by, or merely sitting on the beach in the evening and watching the sunset. The cousins got to know each other a little better (three lived in Colorado and two in Wisconsin), and it brought our whole family closer together. Being near a lake made all the difference, the magic of water influencing people in ways they mostly couldn't express but could surely feel.

While we were at the lake during the summer of 2010, we watched a thunderstorm boil up in the west on a hot July afternoon. By early evening, flashes of lightning cut across the ever-darkening sky, and soon we could hear the rumble of thunder as the storm moved closer. We had gotten everyone out of the boats and off the swimming raft.

When the first drops of rain fell, we moved inside the biggest of our two cabins. Rain splattered against the windows, and then an enormous flash of light with a simultaneous ear-splitting crash of thunder occurred. The lights in our cabin went out, as did those in the other cabins we could see. Lightning had struck a tree less than one hundred yards from the cabin, ripping a gouge in it from top to bottom. We were reminded of the great power of a thunderstorm and why lightning, although awesome to watch, can also be dangerous.

A power outage was a new experience for the grandkids. We found some candles and spent the evening telling stories by candlelight. The grandkids still talk about it when we gather each year: "Do you remember the time lightning struck the tree and power went out?" I must say, the grandkids have become great storytellers, for each year the story is embellished a bit more. I smile when I hear it told.

We still gather at a cabin on a lake every year, these days closer to Madison, where Sue and Steve live with their families. Several of the grandkids are young adults now, and they will drop in and stay for a couple days and then go back to their jobs. Now our great-grandchildren are learning what it is like to spend some time by a lake, and all that being close to a body of water can do for a person.

This yearly gathering of the family, going back thirty-five years, is one of several ways that we have kept our family close and helped the new generations learn what fun it is to live for a few days near a lake.

# Final Thoughts

When my father died at age ninety-three, one way I mourned his passing was to sit by the pond at my farm, alone, for few hours a day. I did nothing but stare at the water, watching the occasional turtle floating by, observing a pair of mallard ducks, looking at the water lilies that were in bloom, thinking and not thinking.

One day during one of my visits, it was raining—just a little, but enough to make that subtle, tinkling sound of rainwater on a pond and to form little round pools on the surface of the still water. Each drop created a small splash and then a circle that expanded and expanded until it disappeared, to be replaced by another. Circles upon circles that came and disappeared.

It was something like life itself. We hope to make a little splash (some hope for larger splashes), with expanding circles that influence the people in our lives, perhaps even beyond. For some, that circle of influence continues after their passing. My father never hoped to make a big splash; it was not his way. His favorite saying was, "Do the best you can with what you've got." That's what he did his entire life. I think he would be pleased that many of his ideas about the importance of water and taking care of the land have been passed down to the generations that followed him.

Sitting by my pond on that warm July afternoon, I thought about how water had influenced all of Pa's life—how he always hoped for rain for his field crops, how he trusted his well to not go dry, how he enjoyed fishing. I had asked him once, after he had turned ninety, what he wished he had done differently in his life. He paused for a moment and then said, "I wish I had gone fishing more often."

With his passing, I thought of thunderstorms and rainy days on the farm, listening to the rain tapping on the barn roof on a June afternoon, and walking behind our small herd of Holsteins dripping with rain on a stormy morning. I remembered the sweet smell of rain after days of dry weather and how all living creatures celebrated the rain's arrival. I remembered fierce downpours that dropped inches of rain and ripped gullies into the sides of the hills on our sandy farm. I recalled walking in the rain to school along a muddy country road, staying inside during recess because it was still raining, digging out the handful of rainy-day board games stacked in the cupboard alongside the woodstove that heated the place and playing checkers with my friend Jim Kolka because we couldn't go outside and play anti-I-over, run sheep run, kick the can, or softball.

Having grown up as I did, helping my family on the farm and living immersed in nature, I learned at an early age to appreciate and respect water. Water supports life, but it can also destroy it, with rampaging floods, mudslides, and soil-depleting erosion. Water can carry life-threatening disease organisms as well as contaminants such as insecticides, farm fertilizer runoff, and animal refuse that can pollute lakes and streams and have long-term detrimental effects on human health and well-being.

Water is often in the news these days. It seems there is either too much, causing floods and loss of life and property, or not enough, making life tenuous for crops, animals, and people. Too often, lakes or rivers become polluted when an agricultural manure storage system leaks, a train hauling petroleum derails, or a river barge loaded with fertilizer hits an obstruction and spills its cargo. The effect on nearby plants, animals, and humans can be devastating.

Irrigation for agriculture increases the demand on water, in some cases depleting nearby aquifers. In addition, fertilizers, pesticides, and herbicides can leach into aquifers, making the water unsuitable for drinking and other uses. Excessive amounts of manure spread on certain soils can also pollute underground aquifers. Hydraulic fracturing in the oil and natural gas industry and its associated sand-mining operations use tremendous amounts of water, sometimes also polluting it with noxious chemicals. Some lakes and rivers have become polluted with heavy metals such as mercury dumped into rivers as industrial waste, making the fish from these waters questionable as food.

Many people take water for granted. Most of us in this country can turn on the tap and fresh, clean water flows out. But those things we take for granted are often in the most danger. I grew up understanding water's importance in the lives of my family and neighbors and learning to conserve it. But for generations of people who did not grow up as I did, water was always just "there," available in such abundance that millions of gallons could be sprayed on ever-thirsty farm crops, urban golf courses, and home lawns so they would be forever green.

For many years people seemed to believe that water was a never-ending resource, especially those living in the Midwest with its thousands of freshwater rivers and lakes. But this belief is slowly changing as people realize that fresh water is not in infinite supply and that it must be cherished and protected along with land and air. By 2016 California had experienced four consecutive years of drought, hurting the state's agriculture industry, limiting the drinking water supply, and threatening wildlife. California's troubles have brought attention to a much larger drought problem throughout the Colorado River basin, and to the nation's water use policies in general.

The Midwest has been blessed with a considerable proportion of the world's freshwater. The Great Lakes—Superior, Michigan, Ontario, Huron, and Erie—contain 84 percent of North America's surface freshwater and about 21 percent of the world's supply of freshwater. Only the polar ice caps have more. But what we have is all we have. As the world's population grows, so does the demand for food, and increased agriculture to meet those demands will require increasing amounts of water. Water for personal and recreational use will also increase with population growth.

The National Resources Defense Council (www.nrdc.org/water) offers the following recommendations to assure an ample supply of freshwater in the future:

1. Promote water efficiency to decrease the amount of water wasted.
2. Protect water from pollution.
3. Ensure that rivers and lakes and associated wetlands have enough water to support aquatic ecosystems.
4. Prepare for the water-related challenges associated with climate change.

We don't know yet precisely how climate change will affect our country's water supply, but the National Resources Defense Council predicts that more than one-third of all counties in the lower forty-eight states face risk of water shortages by 2050. Sea levels are predicted to rise, and intense storms with associated torrential rains will become more prevalent. Thus we will face the contradictory possibility of extreme drought in some places and excessive water in others.

We are at a time in our planet's history when we must realize that we can no longer take water for granted, that it might not always be available to us, fresh and clean, whenever we need it. As a water-dependent society, we should follow the recommendations advocated by the National Defense Council and other concerned groups. But we must go beyond these recommendations and change the fundamental ways we *think* about water.

Two dramatically different belief systems have long existed in this country: those held by indigenous people and those held by the first Europeans who settled here. Indigenous people have long held a spiritual view of water, encompassing the belief that human beings are a part of nature, not apart from it. The predominant European view held by those settling in this country was economic: water is seen as a commodity with potential economic benefits.

To learn to revere, respect, and care for water, we must change our value system from one based purely on economics to one that combines spiritual and practical perspectives. We must step back and examine what it is that we deem important in our lives and for our future. As Aldo Leopold wrote in his seminal work *A Sand County Almanac,* "No important change in ethics was ever accomplished without an internal change in our intellectual emphasis, loyalties, affections and convictions."

As I write this, I think back to the days on our droughty central Wisconsin farm. I know my dad had a great reverence for water, even though he never talked about it that way. His oft-repeated words "Never curse the rain" told me so much more than the words themselves about how he valued and respected water. I've tried to follow in his footsteps.

# Acknowledgments

As a writer I learned early on how to depend on others for help. This book is no exception. I especially want to thank my wife, Ruth, who reads all of my material, and as I often say, if it doesn't get past Ruth, the writing goes nowhere. My longtime editor, Kate Thompson with the Wisconsin Historical Society Press, deserves huge credit and my never-ending thanks for sorting through my words, crossing out some, rearranging others, and ultimately making my work readable. I also want to thank Wisconsin Public TV and especially Jon Miskowski, director of television, and Mik Derks, producer, for seeing this book as source material for an hourlong documentary with the same name.

# About the Author

PHOTO BY STEVE APPS

Jerry Apps is professor emeritus at the University of Wisconsin–Madison and the author of many books on rural history, country life, and the environment. For ten years he wrote a weekly column on nature appreciation for several central Wisconsin newspapers. He has created four documentaries with Wisconsin Public Television and has won several awards for his writing and a regional Emmy Award for the TV program *A Farm Winter.* Jerry and his wife, Ruth, have three children, seven grandchildren, and two great-grandsons. They divide their time between their home in Madison and their farm, Roshara, in Waushara County.

# Discover more books by Jerry Apps

*The Quiet Season: Remembering Country Winters*

*Whispers and Shadows: A Naturalist's Memoir*

*Roshara Journal: Chronicling Four Seasons, Fifty Years, and 120 Acres*

*Garden Wisdom: Lessons Learned from 60 Years of Gardening*

*Old Farm: A History*